AF598093

Methods in Molecular Biology™

Series Editor
John M. Walker
School of Life Sciences
University of Hertfordshire
Hatfield, Hertfordshire, AL10 9AB, UK

For further volumes:
http://www.springer.com/series/7651

Neuronal Cell Culture

Methods and Protocols

Edited by

Shohreh Amini

Department of Biology, College of Science & Technology, Temple University, Philadelphia, PA, USA

Martyn K. White

Department of Neuroscience, Temple University School of Medicine, Philadelphia, PA, USA

Editors
Shohreh Amini
Department of Biology
College of Science & Technology
Temple University
Philadelphia, PA, USA

Martyn K. White
Department of Neuroscience
Temple University
School of Medicine
Philadelphia, PA, USA

ISSN 1064-3745 ISSN 1940-6029 (electronic)
ISBN 978-1-62703-639-9 ISBN 978-1-62703-640-5 (eBook)
DOI 10.1007/978-1-62703-640-5
Springer New York Heidelberg Dordrecht London

Library of Congress Control Number: 2013945403

Cover illustration: Human primary neurons showing synaptodendritic connections. green: immunolabel for axons, red: immunolabel for dendrites. Photo credit: Dianne Langford, William Yen, Yonggang Zhang.

Printed on acid-free paper

Humana Press is a brand of Springer
Springer is part of Springer Science+Business Media (www.springer.com)

Preface

The impressive progress in the field of neuroscience in recent years and the remaining importance of neurological diseases, such as neuro-AIDS, MS, and Alzheimer's, highlight the value of developing methods to grow neural cells and tissues in culture. In this book, we endeavor to provide an overview of the latest aspects of the culture of neural cells while explaining the practical and theoretical considerations of the techniques involved. Starting with a general overview of the neuronal culturing principles that are described, we will cover cell line models for neural cells, the isolation and propagation of primary cultures, stem cells, transfection and transduction of neural cultures, and other more advanced techniques. This book will be of interest to scientists at all levels developing cell culture models for neuroscientific studies.

Philadelphia, PA

Shohreh Amini
Martyn K. White

Contents

Contributors

SHOHREH AMINI • *Department of Biology, College of Science & Technology, Temple University, Philadelphia, PA, USA*
S. AUSIM AZIZI • *Department of Neurology, Temple University School of Medicine, Philadelphia, PA, USA*
JOSEPH F. BONNER • *Department of Neurobiology and Anatomy, Drexel University College of Medicine, Philadelphia, PA, USA*
ARMINE DARBINYAN • *Department of Pathology, The Mount Sinai Medical Center, New York, NY, USA*
NUNE DARBINIAN • *Department of Neuroscience, Temple University School of Medicine, Philadelphia, PA, USA*
PRASUN K. DATTA • *Department of Neuroscience, Temple University School of Medicine, Philadelphia, PA, USA*
LUIS DEL VALLE • *Department of Pathology, Stanley S. Scott Cancer Center, LSU Health Science Center, New Orleans, LA, USA*
KALEN R. DIONNE • *Department of Neurology, Denver Veterans Affairs Medical Center, Aurora, CO, USA*
ITZHAK FISCHER • *Department of Neurobiology and Anatomy, Drexel University College of Medicine, Philadelphia, PA, USA*
JENNIFER GORDON • *Department of Neuroscience, Temple University School of Medicine, Philadelphia, PA, USA*
CHRISTOPHER J. HAAS • *Department of Neurobiology and Anatomy, Drexel University College of Medicine, Philadelphia, PA, USA*
JEE W. HONG • *Shriners Hospital Pediatric Research Center, Temple University School of Medicine, Philadelphia, PA, USA*
WENHUI HU • *Department of Neuroscience, Temple University School of Medicine, Philadelphia, PA, USA*
RAFAL KAMINSKI • *Department of Neuroscience, Temple University School of Medicine, Philadelphia, PA, USA*
KAMEL KHALILI • *Department of Neuroscience, Temple University School of Medicine, Philadelphia, PA, USA*
JANE KOVALEVICH • *Department of Neuroscience, Temple University School of Medicine, Philadelphia, PA, USA*
BARBARA KRYNSKA • *Department of Neurology, Temple University School of Medicine, Philadelphia, PA, USA; Shriners Hospitals Pediatric Research Center, Center for Neural Repair and Rehabilitation, Temple University School of Medicine, Philadelphia, PA, USA*
DIANNE LANGFORD • *Department of Neuroscience, Temple University School of Medicine, Philadelphia, PA, USA*
YINGPENG LIU • *Shriners Hospital Pediatric Research Center, Temple University School of Medicine, Philadelphia, PA, USA*

Paul Pozniak • *Department of Neuroscience, Temple University School of Medicine, Philadelphia, PA, USA*
Ilker Kudret Sariyer • *Department of Neuroscience, Temple University School of Medicine, Philadelphia, PA, USA*
George M. Smith • *Shriners Hospital Pediatric Research Center, Temple University School of Medicine, Philadelphia, PA, USA*
Amanda Parker Struckhoff • *Department of Pathology, Stanley S. Scott Cancer Center, LSU Health Science Center, New Orleans, LA, USA*
Kenneth L. Tyler • *Department of Neurology, Denver Veterans Affairs Medical Center, Aurora, CO, USA*
Martyn K. White • *Department of Neuroscience, Temple University School of Medicine, Philadelphia, PA, USA*
Baheru Woldemichaele • *Department of Neuroscience, Temple University School of Medicine, Philadelphia, PA, USA*
Hassen S. Wollebo • *Department of Neuroscience, Temple University School of Medicine, Philadelphia, PA, USA*
Yonggang Zhang • *Department of Neuroscience, Temple University School of Medicine, Philadelphia, PA, USA*

Chapter 1

General Overview of Neuronal Cell Culture

Jennifer Gordon, Shohreh Amini, and Martyn K. White

Abstract

In this introductory chapter, we provide a general overview of neuronal cell culture. This is a rapidly evolving area of research and we provide an outline and contextual framework for the different chapters of this book. These chapters were all contributed by scientists actively working in the field who are currently using state-of-the-art techniques to advance our understanding of the molecular and cellular biology of the central nervous system. Each chapter provides detailed descriptions and experimental protocols for a variety of techniques ranging in scope from basic neuronal cell line culturing to advanced and specialized methods.

Key words Neurons, Cell culture, Cell lines, Primary cultures, Stem cells, Progenitor cells

The ability to produce in vitro cultures of neuronal cells has been fundamental to advancing our understanding of the functioning of the nervous system. The culture of neuronal cells is particularly challenging since mature neurons do not undergo cell division. One way to overcome this is to establish secondary cell lines that are derived from neuronal tumors and have become immortalized. These have the advantage of being able to be grown fairly easily in cell culture to give unlimited cell numbers as well as minimizing variability between cultures. The disadvantage of these cell lines is that they will show many important physiological differences with the cell type from which they were derived. Often such cell lines are induced to display a more neuronal phenotype by manipulations of the culture conditions, e.g., addition of specific growth factors. In Chapter 2, "Considerations for the Use of SH-SY5Y Neuroblastoma Cells in Neurobiology" by Jane Kovalevich and Dianne Langford, one such cell line is discussed, SH-SY5Y, which has been cultured and used extensively in research on neuronal cells. As described in detail by Kovalevich and Langford, the human SH-SY5Y cell line was derived by subcloning from the parental metastatic bone tumor biopsy cell line SK-N-SH by June Biedler in the 1970s [1]. A modified protocol is presented based on the original culturing conditions described by Biedler for SH-SY5Y cells [2]. The normal

Shohreh Amini and Martyn K. White (eds.), *Neuronal Cell Culture: Methods and Protocols*, Methods in Molecular Biology, vol. 1078, DOI 10.1007/978-1-62703-640-5_1, © Springer Science+Business Media New York 2013

culture conditions for SH-SY5Y cells that are routinely used today remain largely unchanged since the 1970s. SH-SY5Y cells can grow continuously as undifferentiated cells that have a neuroblast-like morphology and express immature but not mature neuronal markers [3]. Differentiated SH-SY5Y cells are morphologically similar to primary neurons with long processes and exhibit a decrease in proliferation rate, exit the cell cycle, and enter G_0, and show increased expression of neuron-specific markers [3]. As described in Chapter 2, a number of agents can be used to induce differentiation, e.g., retinoic acid, phorbol esters, dibutyryl cAMP. Markers for differentiation include βIII-tubulin, synaptophysin, microtubule-associated protein-2 (MAP2), neuron-specific enolase, synaptic associated protein-97 (SAP-97), and neuronal specific nuclear protein NeuN. The morphologies of undifferentiated and differentiated SH-SY5Y cells are shown in Figs. 2.1, 2.2, and 2.3.

Other secondary cell lines have also been used as models to study neuronal cells. In Chapter 3, "Cultured Cell Line Models of Neuronal Differentiation: NT2, PC12," Nune Darbinian describes the culture of NT2 and PC12 cells. NT2, also called NTera, is a human neuronally committed teratocarcinoma cell line that is able to be induced into neuronal cultures by treatment with retinoic acid and inhibitors of mitosis, whereupon they show expression of neuronal markers. Available from the ATCC, NTERA-2 CL.D1 [NT2/D1] (ATCC® CRL-1973™) cells derive from a malignant pluripotent embryonal carcinoma arising in the testis of a 20-year-old male [4]. PC12 is a rat cell line that was derived from a pheochromocytoma of the adrenal medulla [5]. PC12 cells cease to proliferate and undergo terminal differentiation into a neuronal phenotype when treated with nerve growth factor [6]. This cell line is also available from the ATCC—PC-12 (ATCC® CRL-1721™). NT2 and PC12 are suitable host cells for DNA transfection and protein transduction as shown in Figs. 3.4, 3.5, and 3.6. Another type of neuronal cell culture can be derived from mouse teratocarinomas as described in Chapter 4, "Murine Teratocarcinoma-Derived Neuronal Cultures" by Prasun Datta. This chapter describes the culture and propagation of the murine embryonic stem cells, F9 and P19 and strategies for the differentiation of these stem cells into neurons. These include protocols which focus on obtaining enriched populations of mature neurons from P19 cells and differentiation of F9 cells into serotonergic or catecholaminergic neurons. These protocols can be employed to dissect pathways such as gliogenesis and neurogenesis that are involved in differentiation of F9 and P19 into glial cells or terminally differentiated neurons.

While neuronal cell lines have been very useful in the study of neuronal cell cultures and continue to be used today, the use of primary cultures is desirable because they are not tumor-derived and hence are more likely to recapitulate the properties of neuronal cells in vivo. However, unlike cell lines that provide unlimited supplies of homogeneous cells, the preparation and culture of primary cells is much more challenging and this is especially true for neuronal cells. Primary cell cultures are not immortal and hence the number of cells available for experiments is much more limited. Furthermore, since animal tissues in vivo are made up of several different types of cell, it is necessary to separate the cell type of interest from other cell types and to determine the purity of the resulting cultures, e.g., by immunocytochemistry with cell lineage-specific markers. In the case of primary neuronal cell cultures, it is necessary to separate them, as much as possible, from astrocytes and oligodendrocytes. Also with primary cultures, there are other important considerations, i.e., obtaining necessary ethical protocol approvals, which in US universities for example, would come from an Institutional Animal Care and Use Committee for animal cells and from an Institutional Review Board for the protection of human subjects for human tissue. Lastly, it should be noted that primary cultures in general are less easy to transfect than cell lines but that specialized transfection protocols or viral transduction can be used to introduce DNA into these cells, as described later. In Chapter 5, "Isolation and Propagation of Primary Human and Rodent Embryonic Neural Progenitor Cells and Cortical Neurons," Armine Darbinyan and her colleagues describe the preparation and culture of primary neuronal cell cultures and also include culturing of neural and oligodendrocyte progenitor cells.

The brain is not the only source from which neurons can be cultured. Yonggang Zhang and Wenhui Hu describe protocols for the isolation and culture of neurons from the gut of the mouse in Chapter 6, "Mouse Enteric Neuronal Cell Culture." A complex network of neurons embedded within the wall of the gut controls the gastrointestinal tract, which forms the enteric nervous system, also dubbed "the second brain." The enteric nervous system also contains glial cells and consists of at least two plexuses, the myenteric plexus and the submucosal plexus. The isolation of enteric neurons presents technical difficulties not encountered in the isolation of neurons from the brain. There are established protocols for the isolation of enteric neurons from the guinea pig, rat, and human but the mouse is particularly attractive because of the availability of inbred and genetically engineered strains and the economic cost. Zhang and Hu describe two methods to obtain enteric neurons from the mouse myenteric plexuses: (1) direct culture of primary neurons as shown in Fig. 6.1; (2) induction of neuronal differentiation of enteric neural stem/progenitor cells as shown in Fig. 6.2.

Perhaps no more exciting area in cell biology has emerged in recent years than stem cells. This is particularly the case for neuronal cell culture. As noted above, one problem in culturing of neuronal cells is that mature neurons do not proliferate. This is not the case for neural stem cells and they proliferate and are capable of long-term self-renewal. In Chapter 7, "Preparation of Neural Stem Cells and Progenitors: Neuronal Production and Grafting Applications" by Joseph Bonner, Christopher Haas, and Itzhak Fischer, a comprehensive description is given of the techniques involved in the preparation and culture of the stem cells that are involved in the generation of differentiated neurons. This chapter provides in detail the methods required for the isolation, propagation, storage, and differentiation of neural stem cells (NSC) and neural precursor cells (NPC) isolated from rat fetal spinal cords for subsequent in vitro or in vivo studies. Of particular note, Subheading 1.2 provides some useful definitions of terms used in the field, e.g., pluripotency, multipotency, progenitor, precursor, which are all clearly defined. Two major protocols are described in Chapter 7: (1) preparation of Neuroepithelial cells (NEPs) from E10.5 rat spinal cord; (2) preparation of Neuronal Restricted Progenitors (NRPs) and Glial Restricted Progenitor (GRPs) from E13.5 rat spinal cord. NEPs are multipotent cells that can self renew in vivo and can differentiate into neurons, astrocytes, and oligodendrocytes. Thus, NEPs are a true NSC population that represents an intermediate stage between pluripotent immature cells, such as embryonic stem cells and mature differentiated neural cell phenotypes. On the other hand, NRPs/GRPs are a mixed culture of two cell types that together can produce neurons, astrocytes, and oligodendrocytes, but NRPs will only differentiate into neurons while GRPs will only differentiate into astrocytes and oligodendrocytes. Thus, NRPs/GRPs are not a true NSC population but represent a necessary intermediate step between multipotent NSCs and mature neural or glial phenotypes.

In Chapter 8, "Derivation of Neuronal Cells from Fetal Normal Human Astrocyte (NHA)" by Ausim Azizi and Barbara Krynska, protocols are described for a different approach to the differentiation of cells into the neuronal lineage. The starting point here is commercially available cultures of normal human astrocytes (NHA) that have been isolated from normal fetal human brain tissue. Some of the glial fibrillary acidic protein (GFAP)-positive astrocytes have stem cell properties and the genesis of neuronal lineage cells from NHA in adherent culture can be induced by removal of serum and addition of basic fibroblast growth factor (bFGF). These neuronal precursor cells express doublecortin, nestin and are negative for GFAP and can later mature into neurons after withdrawal of the bFGF. Since this model system of neurogenesis is an in vitro system containing of both neurons and glia, it may be thought of as a "human brain in a dish," which is useful for certain studies requiring assays of the effects of various treatments on developing human neurons.

Rather than dispersing cells from a tissue and seeding the cells into primary culture, an alternative is to place unseparated sections of tissue into culture, a technique known as ex vivo culturing or slice culturing. Kalen Dionne and Kenneth Tyler have authored Chapter 9, "Slice Culture Modeling of CNS Viral Infection," which describes this technique applied to the brain. Since the complexity of the central nervous system (CNS) cannot be recapitulated in cultures of disassociated primary cells, the advantage of thin slicing and subsequent culture of CNS tissue is that it allows a valuable means of studying neuronal/glial biology within a physiologically relevant tissue context. Slice culturing facilitates investigator access to both the tissue and the culture medium and these cultures are viable for as long as several weeks after isolation. The protocol described by Dionne and Tyler is based on the membrane-interface method of brain slice culture [7]. Briefly, this procedure involves the placement of explanted rodent brain slices upon a semiporous membrane insert, which sits in a well containing medium such that the slices are suspended at the interface between medium and a humidified atmosphere of 5 % CO_2 at 36 °C (*see* Fig. 9.1). A thin film of medium forms above the slices by capillary action, thus allowing not only hydration and nutrition but also gaseous exchange. In this way, viability can be maintained for several weeks [8].

An important consideration in the isolation of primary cultures is the purity of the culture. There are a number of markers that can be used to label cells for lineage-specific markers to reveal the constituents of the resulting cultures. Classic lineage-specific markers for CNS cells include βIII-tubulin for neuronal cells, GFAP for astrocytic cells, O4 for oligodendrocytic cells, and OX-42/CD-11b for microglial cells. In the case of cultures of stem or progenitor cells, lineage-specific markers also exist for the degree of differentiation, e.g., Nestin and SOX2 for neuroprogenitor cells, NG2 for oligodendrocyte precursor cells. The chapters dealing with primary cultures and stem cell cultures contain information about these markers, for example *see* Chapter 7, Subheadings 2.3 and 3.3. In addition, Chapter 10, "Neurospheres and Glial Cell Cultures: Immunocytochemistry for Cell Phenotyping" by Amanda Parker Struckhoff and Luis Del Valle, describes a detailed procedure for the immunophenotyping of cultures from the CNS. In particular, Table 10.1 of Chapter 10 lists specific biomarkers that are available for the different cell types of the CNS. A complementary approach is suggested by Darbinyan et al. (*see* Chapter 5, Subheading 3.4, **step 2**) which uses Q-RT-PCR to detect and quantify expression of mRNAs for lineage-specific markers. This has the advantage of providing precise quantitation of marker expression but does not address what percentage of cells in a given culture is positive for a marker and hence it should be used in addition to, and not instead of, immunohistochemistry.

Also included are chapters describing common techniques that are employed with neuronal cultures. For example, it is often of interest experimentally to investigate the effects of ectopic expression of a particular protein in a culture. As noted above, primary cultures in general, and neurons in particular, tend to be refractory to transfection. In Chapter 11, "Transfection of Neuronal Cultures," Ilker Sariyer shares his experience of performing transfection experiments with neuronal cells using a variety of different transfection reagents to define an optimized protocol, which is described in detail. As shown in Fig. 11.1, transfection of neuronal cells using this protocol with an expression plasmid for enhanced green fluorescent protein as an indicator for transfected cells, a high efficiency of transfection is achievable. An alternative approach is to perform transduction of the DNA of interest into cells by cloning into a viral vector, packaging the virus and then infecting the target primary culture. In Chapter 11, "Lentiviral Transduction of Neuronal Cells," Hassen Wollebo and coworkers describe the use of a lentivirus vector system for this purpose.

One of the major defining features of neuronal cells is the polarized transmission of information through axons and dendrites. Two compartment systems have now been developed that allow the investigation of these vectoral functions in neuronal cell cultures. In Chapter 13, "Compartmentalized Neuronal Cultures," Armine Darbinyan and coworkers describe the *AX*on *I*nvestigation *S*ystem (AXIS™) device which provides an opportunity to orient neuronal outgrowth and spatially isolate neuronal processes from neuronal bodies. AXIS is a slide-mounted microfluidic system with four wells connected by a channel on each side of the device as shown in Fig. 13.1. Channels are connected by microgrooves that do not permit passage of the cell body but allow extension of neurites. This facilitates the performance of experiments on the control of the extension of neurites and on the specific constituents of neurites in neuronal cell cultures.

Throughout the chapters on neuronal cell cultures, a number of common factors have emerged that are crucial in the cultivation of these cells. One of these is the substrate on which the cells are seeded. For example, in order for PC12 cells to undergo neuronal differentiation in response to nerve growth factor (NGF), they must be plated on collagen IV-coated dishes (*see* Chapter 3). Poly-D-lysine coated dishes/slides are used in the growth of primary neuronal cultures (*see* Chapters 5 and 13). Enteric neurons will also attach to poly-D-lysine coated plates but better adherence can be achieved with the poly-D-lysine/fibronectin or poly-D-lysine/laminin coating. However, double coating is expensive and needs 1–2 days for the coating process (*see* Chapter 6). Poly-L-lysine and laminin coated dishes are used in the culture of neural precursor cells (*see* Chapter 7). Matrigel is also used as a substrate in many applications (*see* Chapters 6, 7, and 6). The importance of the substrate in promoting or inhibiting neurite outgrowth in in vitro cultures of rat dorsal root ganglion (DRG) neurons is explored in the protocols described in Chapter 14 "Quantitative Assessment of Neurite Outgrowth Over Growth Promoting or Inhibitory Substrates" by George Smith and coworkers. This is important in the study of axonal growth, neurotrophic dependence, and the structure and function of growth cones.

Equally as important as the substrate is the composition of the culture medium especially with regard to growth factors and serum. For example, in order for PC12 cells to undergo neuronal differentiation, they are treated with NGF in the absence of serum (*see* Chapter 3). Primary human embryonic neural progenitor cells are cultured in neural stem cell medium, which contains basic fibroblast growth factor (bFGF), epidermal growth factor (EGF), and insulin like growth factor-1 (IGF-1) as described in Chapter 5. The protocols for culturing multipotent neuroepithelial cells (NEPs) use culture medium that is supplemented with bFGF (*see* Chapter 7). In Chapter 8, the genesis of neuronal lineage cells from the astrocytes is described by the removal of serum and exposure to bFGF. The serum-free medium and bFGF induce astrocytes to generate neuronal precursors that express doublecortin and nestin but are negative for GFAP.

Finally, it should be noted that the frequency and cell density is of central importance in neuronal cell culture as is the case for most cells.

In this book, the chapters describe a number of important and up-to-date protocols but it is not intended to be comprehensive. For example, the cell line HEK 293 was generated in the 1970s from normal primary human embryonic kidney (HEK) cells with sheared adenovirus 5 DNA in an early example of the technique of transfection by Graham in the laboratory of van der Eb [9]. Originally thought to be derived by transformation of a fibroblastic, endothelial, or epithelial kidney cell, HEK 293 turned out by later analysis to have characteristics of immature neurons, suggesting that the adenovirus transformed a neuronal lineage cell in the original kidney culture [10]. Thus 293 cells express the neurofilament (NF) subunits NF-L, NF-M, NF-H, and many other neuron-specific proteins [10]. Since other independently derived HEK lines also expressed NFs, this suggests that the human adenoviruses preferentially transform human neuronal lineage cells and that HEK 293 is a human neuronal cell line rather than a kidney epithelial cell line [10]. Finally, it should be noted that recent investigations of cellular plasticity led to the recent discovery that induced neuronal (iN) cells can be generated from mouse and human fibroblasts by expression of defined transcription factors such as Ascl1, Brn2, Olig2, Zic1, and Myt1l. These studies have been the subject of an excellent recent review [11].

References

1. Biedler JL, Helson L, Spengler BA (1973) Morphology and growth, tumorigenicity, and cytogenetics of human neuroblastoma cells in continuous culture. Cancer Res 33: 2643–2652
2. Biedler JL, Roffler-Tarlov S, Schachner M, Freedman LS (1978) Multiple neurotransmitter synthesis by human neuroblastoma cell lines and clones. Cancer Res 38:3751–3757
3. Pahlman S, Ruusala AI, Abrahamsson L, Mattsson ME, Esscher T (1984) Retinoic acid-induced differentiation of cultured human neuroblastoma cells: a comparison with phorbolester-induced differentiation. Cell Differ 14:135–144
4. Andrews PW (1988) Human teratocarcinomas. Biochim Biophys Acta 948:17–36
5. Greene LA, Tischler AS (1976) Establishment of a noradrenergic clonal line of rat adrenal pheochromocytoma cells which respond to nerve growth factor. Proc Natl Acad Sci U S A 73:2424–2428
6. Greene LA (1978) Nerve growth factor prevents the death and stimulates the neuronal differentiation of clonal PC12 pheochromocytoma cells in serum-free medium. J Cell Biol 78:747–755
7. Stoppini L, Buchs PA, Muller D (1991) A simple method for organotypic cultures of nervous tissue. J Neurosci Methods 37:173–182
8. Holopainen IE (2005) Organotypic hippocampal slice cultures: a model system to study basic cellular and molecular mechanisms of neuronal cell death, neuroprotection, and synaptic plasticity. Neurochem Res 30: 1521–1528
9. Graham FL, Smiley J, Russell WC, Nairn R (1977) Characteristics of a human cell line transformed by DNA from human adenovirus type 5. J Gen Virol 36:59–74
10. Shaw G, Morse S, Ararat M, Graham FL (2002) Preferential transformation of human neuronal cells by human adenoviruses and the origin of HEK 293 cells. FASEB J 6:869–871
11. Yang N, Ng YH, Pang ZP, Südhof TC, Wernig M (2011) Induced neuronal cells: how to make and define a neuron. Cell Stem Cell 9:517–525

Chapter 2

Considerations for the Use of *SH-SY5Y* Neuroblastoma Cells in Neurobiology

Jane Kovalevich and Dianne Langford

Abstract

The use of primary mammalian neurons derived from embryonic central nervous system tissue is limited by the fact that once terminally differentiated into mature neurons, the cells can no longer be propagated. Transformed neuronal-like cell lines can be used in vitro to overcome this limitation. However, several caveats exist when utilizing cells derived from malignant tumors. In this context, the popular *SH-SY5Y* neuroblastoma cell line and its use in in vitro systems is described. Originally derived from a metastatic bone tumor biopsy, *SH-SY5Y* (ATCC® CRL-2266™) cells are a subline of the parental line *SK-N-SH* (ATCC® HTB-11™). *SK-N-SH* were subcloned three times; first to *SH-SY*, then to *SH-SY5*, and finally to *SH-SY5Y*. *SH-SY5Y* were deposited to the ATCC® in 1970 by June L. Biedler.

Three important characteristics of *SH-SY5Y* cells should be considered when using these cells in in vitro studies. First, cultures include both adherent and floating cells, both types of which are viable. Few studies address the biological significance of the adherent versus floating phenotypes, but most reported studies utilize adherent populations and discard the floating cells during media changes. Second, early studies by Biedler's group indicated that the parental differentiated *SK-N-SH* cells contained two morphologically distinct phenotypes: neuroblast-like cells and epithelial-like cells (Ross et al., J Nat Cancer Inst 71:741–747, 1983). These two phenotypes may correspond to the "N" and "S" types described in later studies in *SH-SY5Y* by Encinas et al. (J Neurochem 75:991–1003, 2000). Cells with neuroblast-like morphology are positive for tyrosine hydroxylase (TH) and dopamine-β-hydroxylase characteristic of catecholaminergic neurons, whereas the epithelial-like counterpart cells lacked these enzymatic activities (Ross et al., J Nat Cancer Inst 71:741–747, 1983). Third, *SH-SY5Y* cells can be differentiated to a more mature neuron-like phenotype that is characterized by neuronal markers. There are several methods to differentiate *SH-SY5Y* cells and are mentioned below. Retinoic acid is the most commonly used means for differentiation and will be addressed in detail.

Key words Neuroblastoma, Differentiation, Neuron

1 Introduction

1.1 Preface

A detailed report by Biedler et al. in 1978 describes the original culturing conditions for *SH-SY5Y* cells [3] and a modified version of this protocol is presented in Subheading 3 of this chapter. In brief, *SK-N-SH* cells and derived clones including *SH-SY5Y* are plated at a density of 2×10^5 or 4×10^6 cells/60 mm dish in Eagle's

Shohreh Amini and Martyn K. White (eds.), *Neuronal Cell Culture: Methods and Protocols*, Methods in Molecular Biology, vol. 1078, DOI 10.1007/978-1-62703-640-5_2, © Springer Science+Business Media New York 2013

minimum essential medium supplemented with nonessential amino acids, 15 % fetal bovine serum, penicillin (100 IU/ml) and streptomycin (100 μg/ml). These original plating densities were used to determine the doubling time and saturation densities. Although *SH-SY5Y* doubling time was not reported specifically, the parental neuroblast-like populations has a doubling time of approximately 27 h and the subclones were reported to have similar doubling times. *SH-SY5Y* cells are reported to have a growth saturation density of $>1 \times 10^6$ cells/cm^2. In this study, both adherent and floating cells were collected when cells were passed. Floating cells were removed in culture medium, whereas adherent cells were detached with trypsin. The two cell populations were combined, centrifuged, and re-plated at appropriate densities. Combining the suspended and adherent cells may become an important aspect during differentiation protocols and will be discussed in later sections. Cells are grown in a humidified chamber with 5 % CO_2 at 37 °C. Little has changed over the past four decades with regard to normal culture conditions for *SH-SY5Y* cells. These early studies also addressed transmitter properties and indicated that *SH-SY5Y* consisted of homogeneous neuroblast-like populations. Analyses of specific neuronal enzyme activities in *SK-N-SH* cells and clones indicated dopamine-β-hydroxylase levels of 3.74 nmol/h/mg in *SH-SH5Y* cells, although only one set of cell cultures was assayed [3]. Levels of choline acetyl-transferase, acetyl-cholinesterase, and butyryl-cholinesterase were negligible in *SK-N-SH* and its clones, including *SH-SY5Y*.

1.2 Undifferentiated Versus Differentiated SH-SY5Y Cells

The ability for researchers to differentiate *SH-SY5Y* neuroblastoma cells into cells possessing a more mature, neuron-like phenotype through manipulation of the culture medium has afforded numerous benefits in the field of neuroscience research. Advantages include the capacity for large-scale expansion prior to differentiation, with relative ease and low cost to culture compared to primary neurons. Since these cells are considered a cell-line, the ethical concerns associated with primary human neuronal culture are not involved. Additionally, since *SH-SY5Y* cells are human-derived, they express a number of human-specific proteins and protein isoforms that would not be inherently present in rodent primary cultures. Furthermore, differentiation synchronizes the cell cycle, which can fluctuate dramatically in undifferentiated *SH-SY5Y* cells and other commonly used cell lines, to produce a homogenous neuronal cell population [1, 2].

Both undifferentiated and differentiated *SH-SY5Y* cells have been utilized for in vitro experiments requiring neuronal-like cells. Neuronal differentiation entails a number of specific events, including formation and extension of neuritic processes, increased electrical excitability of the plasma membrane, formation of synaptophysin-positive functional synapses, and induction of

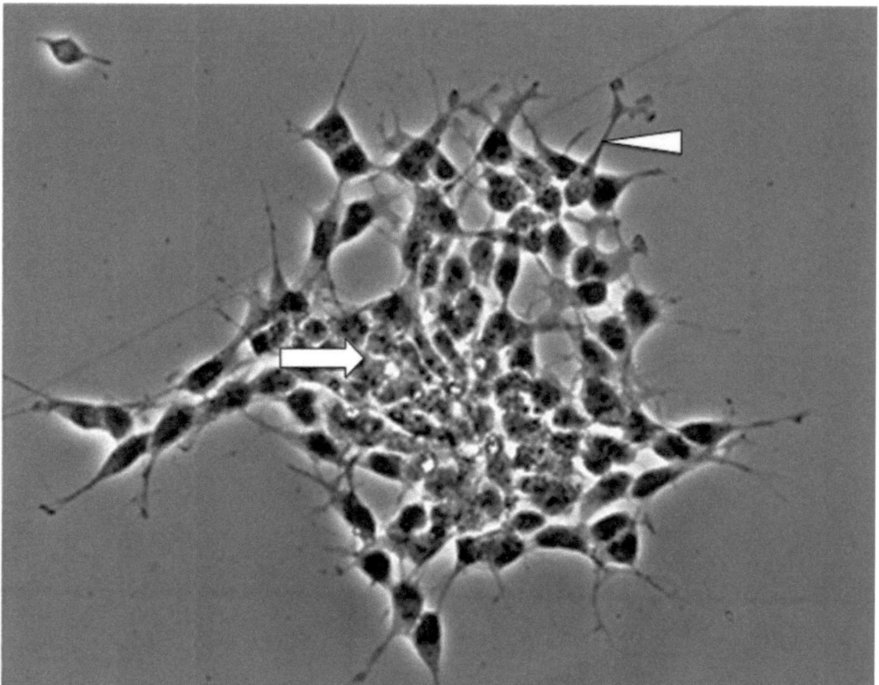

Fig. 1 Undifferentiated *SH-SY5Y* cells. Cells tend to grow in clusters and may form clumps of rounded cells on top of one another (*arrow*). At edges of the cluster, cells begin to extend short neurites (*arrowhead*)

neuron-specific enzymes, neurotransmitters, and neurotransmitter receptors [4–8]. Thus, when determining whether undifferentiated or differentiated cells should be utilized for a particular experiment, all of these properties should be taken into consideration.

In the undifferentiated form, *SH-SY5Y* cells are characterized morphologically by neuroblast-like, non-polarized cell bodies with few, truncated processes. The cells tend to grow in clusters and may form clumps as cells appear to grow on top of one another in the central region of a cell mass (Fig. 1). Likewise, the cultures contain both adherent and floating cells, and some studies suggest that the floating cells are more likely to adhere and differentiate into "N" type cells upon RA-differentiation than the adherent cells present in undifferentiated cultures. Undifferentiated *SH-SY5Y* cells continuously proliferate, express immature neuronal markers, and lack mature neuronal markers [6]. Undifferentiated cells are considered to be most reminiscent of immature catecholaminergic neurons [7, 9]. Following treatment with differentiation-inducing agents, *SH-SY5Y* cells become morphologically more similar to primary neurons with long, exquisite processes [6] (Fig. 2). Mature cells may exhibit numerous but randomly distributed processes or become distinctly polarized, depending on the differentiation induction method. Differentiation of *SH-SY5Y* cells also induces a decrease in proliferation rate, as cells are withdrawn from the cell

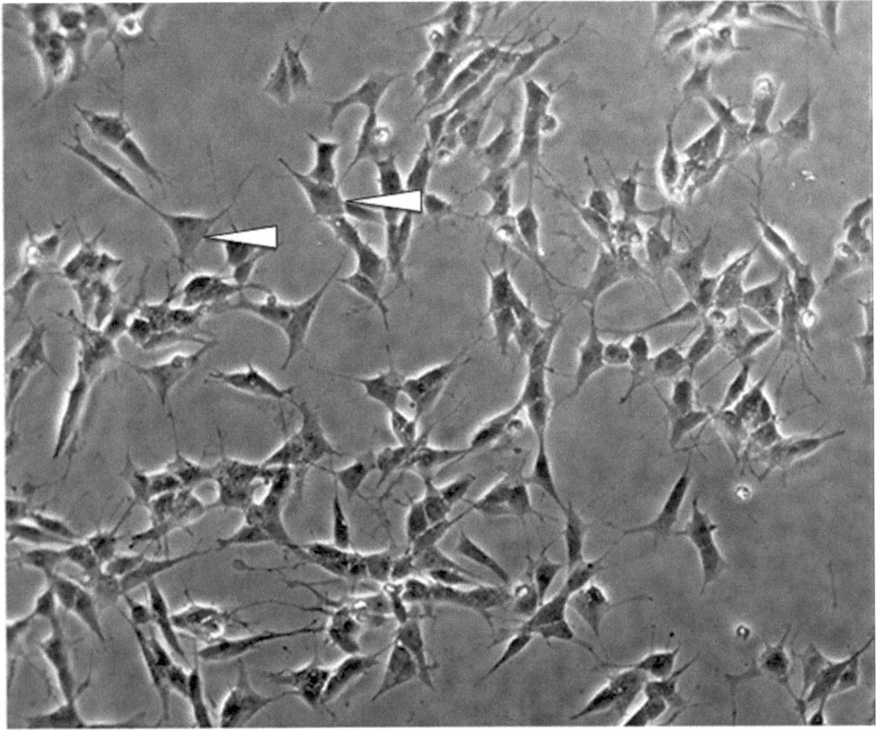

Fig. 2 Differentiated *SH-SY5Y* cells. Cells do not cluster and have a more pyramidal shaped cell body (*arrowhead*). Neurites begin to extend, reminiscent of dendrites and/or axons

cycle, and an increase in the activity of neuron specific enolase (NSE), the dominant enolase-isoenzyme present in neuronal and neuroendocrine tissues [2, 6]. A number of methods exist for induction of differentiation in *SH-SY5Y* cells, and are mentioned below. *SH-SY5Y* cells can be driven toward a variety of adult neuronal phenotypes including cholinergic, adrenergic, or dopaminergic, depending on media conditions [9]. The differentiation method selected for in vitro experiments should ultimately be determined by the desired phenotype following differentiation, as well as for the reduction of non-target effects on experimental pathways in question by particular differentiating agents.

1.3 Retinoic Acid

One of the most commonly implemented and best-characterized methods for induction of differentiation in *SH-SY5Y* cells is through addition of retinoic acid (RA) to the cell culture medium. Retinoic acid is a vitamin A derivative known to possess powerful growth-inhibiting and cellular differentiation-promoting properties [10, 11]. In fact, vitamin A deficiency is linked to the development of squamous metaplasia in various epithelial tissues, while administration of vitamin A can reverse these effects and restore normal cellular differentiation [10, 11]. Typically, RA is administered at a concentration of 10 μM for a minimum of 3–5 days in serum-free or low serum medium to induce differentiation [6, 12], although slight variations in media are reported.

Retinoic acid treatment has been shown to promote survival of *SH-SY5Y* cells through activation of the phosphatidylinositol 3-kinase/Akt signaling pathway and upregulation of the anti-apoptotic Bcl-2 protein [13, 14]. Furthermore, some studies show that RA-differentiated cells are less vulnerable than undifferentiated cells to toxin-mediated cell death induced by agents including 6-hydroxydopamine (6-OHDA), 1-methyl-4-phenyl-1,2,3,6-tetrahydropyridine (MPTP), or its metabolite, 1-methyl-4-phenyl-pyridinium ion (MPP^+) than undifferentiated cells [12].

SH-SY5Y cells differentiate primarily to a cholinergic neuron phenotype in response to RA treatment, as evidenced by increased expression of choline acetyl transferase (ChAT) activity and vesicular monamine transporter (VMAT) expression [7, 15]. Cells may also be driven toward a mature dopaminergic phenotype by RA treatment, but this typically requires co-administration of additional agents such as phorbol esters [15]. While controversy exists in the literature over whether dopaminergic markers present in undifferentiated *SH-SY5Y* cells significantly increase during RA-induced differentiation, studies demonstrate robust increases in tyrosine hydroxylase (TH), dopamine receptor 2 and 3 subtypes (D2R and D3R), and dopamine transporter (DAT) expression when RA administration is followed by treatment with phorbol esters [15]. Interestingly, Encinas et al. reported in 2000 that RA-differentiated *SH-SY5Y* cells responded to carbachol stimulation by increased release of noradrenaline [2]

1.4 Phorbol Esters

In addition to RA-mediated differentiation, *SH-SY5Y* cells have been shown to differentiate in the presence of phorbol esters such as 12-*O*-tetradecanoyl-phorbol-13 acetate (TPA) [5]. In 1981, Påhlman et al. demonstrated that *SH-SY5Y* cells exposed to 1.6×10^{-8} M TPA for 4 days appeared morphologically differentiated with long, straight processes of uneven appearance and frequent varicosities [5]. TPA treatment also resulted in partial growth inhibition, an approximate twofold increase in NSE activity, and the appearance of cytoplasmic neurosecretory granula, which can be visualized by electron microscopy [5, 6]. One striking difference between TPA- and RA-induced differentiation of *SH-SY5Y* cells is that TPA treatment increases the cellular noradrenaline content up to 200-fold while RA treatment causes an approximate fourfold induction in noradrenaline [6]. Therefore, use of TPA to induce differentiation of *SH-SY5Y* cells produces a predominantly adrenergic cellular phenotype [16, 17].

1.5 Dibutyryl Cyclic AMP

Hormones and neurotransmitters that upregulate intracellular levels of cyclic AMP (cAMP) promote differentiation and long-term potentiation in neuronal cells [18, 19]. Exposure of *SH-SY5Y* cells to dibutyryl cyclic AMP (dbcAMP) results in neurite extension, as well as in increased expression of the mature neuronal marker growth-associated protein 43 (GAP43) [19, 20]. Studies demonstrate that treatment with 1 mM dbcAMP for 3 days decreases cell aggregation, reminiscent of the aggregation commonly observed in undifferentiated cultures (Fig. 1), and induces significant neurite elongation and branching [20]. Exposure to dbcAMP also leads to a significant increase in tyrosine hydroxylase (TH) immunoreactivity and in the cellular content of noradrenaline in a protein kinase A (PKA)-dependent manner [14, 20]. In comparison to RA and TPA treatment, which increase expression of Bcl-2, dbcAMP decreases Bcl-2, highlighting one biochemical difference among differentiation methods. These studies indicate that differentiation of *SH-SY5Y* cells with dbcAMP produces a morphological phenotype similar to that seen in RA and TPA-differentiated cells, and that the differentiated culture is comprised of primarily adrenergic neuron-like cells.

1.6 Additional Methods of Differentiation

A number of less-common alternative methods to induce differentiation in *SH-SY5Y* cells have been described, as well. Staurosporine, a PKC inhibitor triggers neuritogenesis and cell cycle arrest in *SH-SY5Y* cells [21, 22]. However, unlike cells differentiated with RA, staurosporine-treated cells display increased vulnerability to toxic insults including cisplatin, 5-fluorouracil, 6-OHDA, and γ-radiation, and express decreased levels of Bcl-2 [22]. Consequently, staurosporine-treated cells undergo apoptosis in a dose-dependent manner. In addition to staurosporine, treatment with growth factors such as nerve growth factor (NGF) and brain-derived neurotrophic factor (BDNF) has been shown to support differentiation into and maintenance of a mature neuronal phenotype, particularly when used in combination with RA or TPA treatment [2, 23, 24]. Similarly, culturing *SH-SY5Y* cells in neurobasal medium with B27 supplement, conditions commonly used for primary neuronal culture, has been shown to enhance differentiation [25]. Other methods include treatment with cholesterol, vitamin D, or insulin, or culturing cells on a substrate designed to promote neuronal differentiation and survival [26–29]. Again, selection of a suitable differentiation induction agent should be carefully evaluated depending on the possibility of downstream, unintended effects on output measures caused by treatment alone.

1.7 Markers for Differentiation

Undifferentiated *SH-SY5Y* cells typically resemble immature catecholaminergic neurons. They are characterized by markers indicative of proliferation, such as proliferating cell nuclear antigen (PCNA), as well as by immature neuronal markers such as nestin [7, 30]. Undifferentiated cells also express differentiation-inhibiting basic helix-loop-helix transcription factors ID1, ID2, and ID3, all of which are significantly decreased following

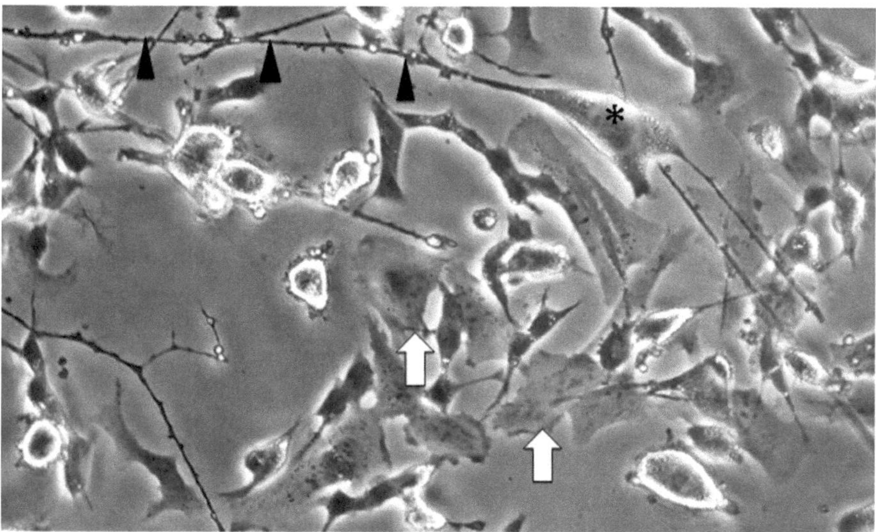

Fig. 3 Differentiated cells populations consist of two morphologically distinct types: "S" and "N". The "S" type cell is epithelial-like with no processes (*arrows*), whereas the "N" type is more neuronal-like with pyramidal shaped bodies (*asterisk*) and long processes (*arrowheads*)

treatment with RA or TPA [13]. Following differentiation, *SH-SY5Y* cells express a number of mature neuronal markers, including βIII-tubulin, microtubule-associated protein-2 (MAP2), synaptophysin, NeuN, synaptic associated protein-97 (SAP-97), and NSE [7, 12]. Additionally, expression of differentiation-promoting genes *NEUROD6* and *NEUROD1* increase following RA treatment [13]. Differentiation of *SH-SY5Y* cells can result in a relatively homogenous population of G0 stage, neuronal-like cells which display an absence of markers for other CNS cell types, such as the astrocytic marker glial fibrillary acidic protein (GFAP) [2, 11], but care should be given to assess proliferation of cells that lack the neuronal-like phenotype as discussed below.

Numerous reports have specifically addressed methods to obtain neuron-like cells from undifferentiated *SH-SY5Y*. Some reports suggest using only floating cells in differentiation protocols, while others recommend using only adherent cells for differentiation. A detailed description for RA-differentiated, BDNF maintained *SH-SY5Y* cells was published in 2000 by Encinas et al. [2]. In these studies, the authors classify undifferentiated cells as either S-type (substrate adherent) without neuron-like phenotype or N-type (neuroblastic) characterized by neuritic processes (Fig. 3). In the study by Encinas, cells were plated at 10^4 cells/cm^2 on collagen-coated (0.05 mg/ml) plates in DMEM with 2 mM L-glutamine, penicillin (20 IU/ml), streptomycin (20 mg/ml) and 15 % FBS. Upon differentiation with 10 μm all-*trans*-RA for 5 days, N-type cells were reported to gain neuron-like characteristics more readily than S-type. However, at day 10 post-RA, the percentage of

S-type cells in the culture increased and cultures were overgrown by the S-type population. Thus, short term RA treatment (up to 5 days) appeared to induce differentiation of N-type cells, but longer treatment (>10 days) promoted proliferation of S-type cells thereby promoting unbalanced proportion of S- to N-type cells. Based on results such as these, consideration should be given not only to passage number, but also to cell phenotype in culture when designing in vitro experiments. One potential solution to prevent the imbalance, recommended by Encinas et al. is adding 50 ng/ml BDNF to RA-treated cultures since RA is reported to induce TrkB receptors on *SH-SY5Y* cells with maximal expression 5 days after BDNF exposure. Removal of serum in combination with BDNF treatment is recommended to prevent replication of S-type cells. Including serum during RA-BDNF treatments can promote S-type proliferation that is observed as a monolayer of S cells under the differentiating N-type cells. Cultures maintained under serum-free, BDNF supplemented conditions were stable for up to 3 weeks without S-type cell over growth with minimal apoptotic cells as determined by TUNEL assay and biochemical analyses. With BDNF cells are arrested, but BDNF withdrawal induces cells to enter the S-phase of the cell cycle [2].

Neuronal marker expression of cultures exposed to RA, BDNF and serum-free conditions was assessed by immunocytochemistry and western analyses. In these studies, neither untreated nor RA-BDNF treated *SH-SHY5Y* cells expressed GFAP, but NSE was detected in all conditions. On the other hand, both medium and high molecular weight neurofilament was detected in untreated and 5-day RA treated neurons, but disappeared after BDNF was added. GAP-43 expression increased after 5 days RA treatment followed by 1 day of BDNF treatment. However, continued BDNF treatment for 3–9 days resulted in return of GAP-43 levels to baseline. The authors suggest that this expression profile coincides with the high levels of neurite extension during which GAP-43 is required. However, the removal of RA from BDNF-containing medium may have contributed, as well.

1.8 Markers for Receptors/Transporters

1.8.1 Dopaminergic Neurons

SH-SY5Y cells in both undifferentiated and differentiated states express a number of dopaminergic neuronal markers. These cells express TH, an enzyme critical for the catalysis of dopamine and, further downstream, noradrenaline (norepinephrine) and adrenaline (epinephrine) [20]. Importantly, *SH-SY5Y* cells also express the dopamine transporter (DAT), as well as dopamine receptor subtypes 2 and 3 (D2R and D3R), making them an exemplary in vitro system for the study of neurotoxicity in dopaminergic neurons as well as for drugs, which are known to produce primary effects through activation of dopamine receptors [31]. Expression of TH, DAT, D2R, and D3R has been shown to increase following differentiation in a number of studies, particularly when a combination of RA and TPA are utilized to induce differentiation [7, 15, 32]. However, there also exists a body of literature that reports no significant difference in levels of TH or DAT following differentiation, although these studies typically utilized RA alone or agents, which promote a non-dopaminergic phenotype to induce differentiation [12, 33].

1.8.2 Adrenergic Neurons

SH-SY5Y cells can be driven toward an adrenergic phenotype by RA- or TPA-induced differentiation. Again, the expression of TH is required, as nor-adrenaline and adrenaline are derived from dopamine catalyzed in a TH-dependent manner [20]. Studies demonstrate that both differentiated and high-passage undifferentiated *SH-SY5Y* cells express sufficient levels of dopamine-β-hydroxylase, the enzyme that catalyzes the formation of nor-adrenaline from dopamine, and are capable of converting intracellular dopamine into nor-adrenaline (nor-epinephrine). Additionally, *SH-SY5Y* cells express the norepinephrine transporter (NET) and the vesicular monoamine transporter (VMAT), characteristic of adrenergic neurons.

1.8.3 Cholinergic Neurons

The expression of both muscarinic and nicotinic acetylcholine receptors has been reported in *SH-SY5Y* cells. The G-protein coupled muscarinic receptors are present on membranes of both undifferentiated and differentiated cells, though levels and binding properties are differentially regulated according to method of differentiation. For instance, TPA treatment decreases while RA treatment increases the number of muscarinic binding sites [4]. Both TPA- and RA-treated cells display significantly increased acetylcholinesterase activity compared to undifferentiated cells. However, only RA treatment has been shown to increase choline acetyltransferase activity [4]. The ligand-gated ion channel nicotinic acetylcholine receptors (nAChR) are also present in *SH-SY5Y* cells and are deemed to be analogous to human ganglia-type nAChR [34, 35]. The nAChR on *SH-SY5Y* cells have been shown to desensitize in response to nicotine but regain full sensitivity following washout, similarly to what is known to happen in primary neurons upon receptor activation [36].

2 Materials

2.1 Cells

Cells can be purchased from ATCC (ATCC® CRL-2266) http://www.atcc.org/. Cells will arrive on dry ice and should be processed immediately or placed in vapor phase liquid nitrogen until ready for culturing.

2.2 Reagents

1. Dulbecco's Minimum Essential Medium (DMEM).
2. F12 Medium.

3. Fetal Bovine Serum (FBS), 10 % final concentration.
4. Penicillin/Streptomycin (Pen/Strep), 1 % final concentration (100 IU/ml, 100 μg/ml, respectively).
5. Trypsin/EDTA.
6. Neurobasal medium (NB).
7. B27 supplement.
8. GlutaMAX.
9. All-*trans*-retinoic acid (ATRA).
10. DMSO.
11. 1× Phosphate-buffered saline (PBS).
12. Sterile disposable filter apparatus, with 0.22 μm pore size.

3 Methods

3.1 Initial Culture Conditions

1. Prepare growth medium: DMEM/F12 (1:1, v:v) medium, (10 % FBS and 1 % pen/strep) and filter through a 0.22 μm pore filter apparatus.
2. Obtain cells from the ATCC (ATCC® CRL-2266) and thaw quickly at 37 °C.
3. Gently remove cell suspension from tube and add to a T-75 tissue culture flask containing warm (37 °C) growth medium (*see* **Note 1**).
4. Culture cells at 37 °C, 5 % CO_2. Growth medium should be refreshed every 4–7 days (*see* **Note 2**). Monitor cells for confluence. When cells reach 80–90 % confluence, subculture as described below.

3.2 Sub-culture Conditions

1. Aspirate medium under sterile conditions. If a number of floating cells are present, medium can be collected and centrifuged so non-adherent cells can be recovered and re-plated.
2. Rinse adherent cells once with sterile 1× PBS pre-warmed to 37 °C or room temperature (*see* **Note 3**). To prevent detaching cells, add the PBS to the inside surface of the culture flask that does not have cells attached. Do not add PBS directly onto cell monolayer. Gently tip the flask so that the PBS washes over the cell monolayer. Aspirate PBS.
3. Add trypsin to adherent cells for approximately 2 min or until cells visibly detach from culture flask (*see* **Note 4**).
4. Neutralize trypsin by adding an equal volume of DMEM/F12 medium containing 10 % FBS.
5. Collect detached cell suspension and centrifuge at 1,500 rpm for 5 min at room temperature to concentrate cell pellet.

6. Aspirate supernatant carefully without disturbing the cell pellet.
7. Gently suspend pellet in DMEM/F12 medium containing 10 % FBS.
8. To separate cells from clumps, pipette up and down gently until the suspension appears homogenous (*see* **Note 1**).
9. Count cells using a hemocytometer and plate at approximately 3×10^3 to 1×10^5 cells/cm^2 (*see* **Note 5**).

3.3 Differentiation

1. 24–48 h after plating, replace serum-containing medium with Neurobasal medium (containing B27 supplement and GlutaMAX) and 10 μM all-*trans*-retinoic acid (ATRA) to promote differentiation and neuronal phenotype.
2. Allow cells to grow in ATRA-containing neurobasal medium for a minimum of 3–5 days, refreshing the medium every 48 h.
3. Differentiation can be monitored microscopically via morphological assessment of neurite outgrowth (compare Figs. 1 and 2).

3.4 Freezing Down

1. To freeze down *SH-SY5Y* cells, begin with undifferentiated cultures.
2. Harvest 80–90 % confluent monolayer from a T75 flask and pellet cells as described above (*see* **Note 6**).
3. Suspend the cell pellet gently in 1 ml of 90 % FBS, 10 % DMSO in a sterile 1.5 ml screw cap vial appropriate for storage in the vapor phase of liquid nitrogen.
4. Store cells at −80 °C for approximately 24 h in an insulated cryobox and then transfer tubes to liquid nitrogen for long-term storage.

4 Notes

1. To improve cell survival, be careful not to introduce air when pipetting/transferring cells between flasks or tubes.
2. The color of the medium indicates the rate of metabolism of key components by cells. When medium becomes more acidic (color appears more yellow than red), it is likely time to change the medium.
3. Rinsing with PBS removes the majority of the serum contained in growth medium. Trypsin works more efficiently in the absence of serum as serum inhibits its activity.
4. Reduce the amount of time cells are exposed to trypsin. Once cells have been exposed to trypsin for about 1 min, you may gently tap the flask to assist in detachment.

5. You will discover that SHSY5Y cells need to be plated at a density conducive to cell–cell communication to proliferate. If cells are plated too sparsely, growth rate is reduced and cell death is high.
6. A T75 flask that is about 89–90 % confluent contains an appropriate number of cells to freeze down in approximately 1 ml of freeze medium allowing you to plate into a T75 flask upon thawing and re-culturing. This is important because when cells are thawed and re-plated after having been in liquid nitrogen, inevitably some cells will die. The density of surviving cells should allow the culture to be maintained at least 24 h before sub-culturing is necessary.

References

1. Ross RA, Spengler BA, Biedler JL (1983) Coordinate morphological and biochemical interconversion of human neuroblastoma cells. J Natl Cancer Inst 71(4):741–747
2. Encinas M, Iglesias M, Liu Y, Wang H, Muhaisen A, Cena V, Gallego C, Comella JX (2000) Sequential treatment of SH-SY5Y cells with retinoic acid and brain-derived neurotrophic factor gives rise to fully differentiated, neurotrophic factor-dependent, human neuron-like cells. J Neurochem 75(3):991–1003
3. Biedler JL, Roffler-Tarlov S, Schachner M, Freedman LS (1978) Multiple neurotransmitter synthesis by human neuroblastoma cell lines and clones. Cancer Res 38(11 Pt 1):3751–3757
4. Adem A, Mattsson ME, Nordberg A, Pahlman S (1987) Muscarinic receptors in human SH-SY5Y neuroblastoma cell line: regulation by phorbol ester and retinoic acid-induced differentiation. Brain Res 430(2):235–242
5. Pahlman S, Odelstad L, Larsson E, Grotte G, Nilsson K (1981) Phenotypic changes of human neuroblastoma cells in culture induced by 12-O-tetradecanoyl-phorbol-13-acetate. Int J Cancer 28(5):583–589
6. Pahlman S, Ruusala AI, Abrahamsson L, Mattsson ME, Esscher T (1984) Retinoic acid-induced differentiation of cultured human neuroblastoma cells: a comparison with phorbolester-induced differentiation. Cell Differ 14(2):135–144
7. Lopes FM, Schroder R, da Frota ML Jr, Zanotto-Filho A, Muller CB, Pires AS, Meurer RT, Colpo GD, Gelain DP, Kapczinski F, Moreira JC, Fernandes Mda C, Klamt F (2010) Comparison between proliferative and neuron-like SH-SY5Y cells as an in vitro model for Parkinson disease studies. Brain Res 1337: 85–94. doi:10.1016/j.brainres.2010.03.102
8. Tosetti P, Taglietti V, Toselli M (1998) Functional changes in potassium conductances of the human neuroblastoma cell line SH-SY5Y during in vitro differentiation. J Neurophysiol 79(2):648–658
9. Xie HR, Hu LS, Li GY (2010) SH-SY5Y human neuroblastoma cell line: in vitro cell model of dopaminergic neurons in Parkinson's disease. Chin Med J 123(8):1086–1092
10. Lotan R (1996) Retinoids in cancer chemoprevention. FASEB J 10(9):1031–1039
11. Melino G, Thiele CJ, Knight RA, Piacentini M (1997) Retinoids and the control of growth/death decisions in human neuroblastoma cell lines. J Neurooncol 31(1–2):65–83
12. Cheung YT, Lau WK, Yu MS, Lai CS, Yeung SC, So KF, Chang RC (2009) Effects of all-trans-retinoic acid on human SH-SY5Y neuroblastoma as in vitro model in neurotoxicity research. Neurotoxicology 30(1):127–135. doi:10.1016/j.neuro.2008.11.001
13. Lopez-Carballo G, Moreno L, Masia S, Perez P, Barettino D (2002) Activation of the phosphatidylinositol 3-kinase/Akt signaling pathway by retinoic acid is required for neural differentiation of SH-SY5Y human neuroblastoma cells. J Biol Chem 277(28):25297–25304. doi:10.1074/jbc.M201869200
14. Itano Y, Ito A, Uehara T, Nomura Y (1996) Regulation of Bcl-2 protein expression in human neuroblastoma SH-SY5Y cells: positive and negative effects of protein kinases C and A, respectively. J Neurochem 67(1):131–137
15. Presgraves SP, Ahmed T, Borwege S, Joyce JN (2004) Terminally differentiated SH-SY5Y cells provide a model system for studying neuroprotective effects of dopamine agonists. Neurotox Res 5(8):579–598
16. Murphy NP, Ball SG, Vaughan PF (1991) Potassium- and carbachol-evoked release of

[3H]noradrenaline from human neuroblastoma cells, SH-SY5Y. J Neurochem 56(5): 1810–1815

17. Scott IG, Akerman KE, Heikkila JE, Kaila K, Andersson LC (1986) Development of a neural phenotype in differentiating ganglion cell-derived human neuroblastoma cells. J Cell Physiol 128(2):285–292. doi:10.1002/jcp.1041280221
18. Frey U, Huang YY, Kandel ER (1993) Effects of cAMP simulate a late stage of LTP in hippocampal CA1 neurons. Science 260(5114):1661–1664
19. Sanchez S, Jimenez C, Carrera AC, Diaz-Nido J, Avila J, Wandosell F (2004) A cAMP-activated pathway, including PKA and PI3K, regulates neuronal differentiation. Neurochem Int 44(4):231–242
20. Kume T, Kawato Y, Osakada F, Izumi Y, Katsuki H, Nakagawa T, Kaneko S, Niidome T, Takada-Takatori Y, Akaike A (2008) Dibutyryl cyclic AMP induces differentiation of human neuroblastoma SH-SY5Y cells into a noradrenergic phenotype. Neurosci Lett 443(3):199–203. doi:10.1016/j.neulet.2008.07.079
21. Leli U, Shea TB, Cataldo A, Hauser G, Grynspan F, Beermann ML, Liepkalns VA, Nixon RA, Parker PJ (1993) Differential expression and subcellular localization of protein kinase C alpha, beta, gamma, delta, and epsilon isoforms in SH-SY5Y neuroblastoma cells: modifications during differentiation. J Neurochem 60(1):289–298
22. Tieu K, Zuo DM, Yu PH (1999) Differential effects of staurosporine and retinoic acid on the vulnerability of the SH-SY5Y neuroblastoma cells: involvement of bcl-2 and p53 proteins. J Neurosci Res 58(3):426–435
23. Chen J, Chattopadhyay B, Venkatakrishnan G, Ross AH (1990) Nerve growth factor-induced differentiation of human neuroblastoma and neuroepithelioma cell lines. Cell Growth Differ 1(2):79–85
24. Simpson PB, Bacha JI, Palfreyman EL, Woollacott AJ, McKernan RM, Kerby J (2001) Retinoic acid evoked-differentiation of neuroblastoma cells predominates over growth factor stimulation: an automated image capture and quantitation approach to neuritogenesis. Anal Biochem 298(2):163–169. doi:10.1006/abio.2001.5346
25. Sayas CL, Moreno-Flores MT, Avila J, Wandosell F (1999) The neurite retraction induced by lysophosphatidic acid increases Alzheimer's disease-like Tau phosphorylation. J Biol Chem 274(52):37046–37052
26. Sarkanen JR, Nykky J, Siikanen J, Selinummi J, Ylikomi T, Jalonen TO (2007) Cholesterol supports the retinoic acid-induced synaptic vesicle formation in differentiating human SH-SY5Y neuroblastoma cells. J Neurochem 102(6):1941–1952. doi:10.1111/j.1471-4159.2007.04676.x
27. Recio-Pinto E, Ishii DN (1984) Effects of insulin, insulin-like growth factor-II and nerve growth factor on neurite outgrowth in cultured human neuroblastoma cells. Brain Res 302(2):323–334
28. Reddy CD, Patti R, Guttapalli A, Maris JM, Yanamandra N, Rachamallu A, Sutton LN, Phillips PC, Posner GH (2006) Anticancer effects of the novel 1alpha, 25-dihydroxyvitamin D3 hybrid analog QW1624F2-2 in human neuroblastoma. J Cell Biochem 97(1):198–206. doi:10.1002/jcb.20629
29. Agholme L, Lindstrom T, Kagedal K, Marcusson J, Hallbeck M (2010) An in vitro model for neuroscience: differentiation of SH-SY5Y cells into cells with morphological and biochemical characteristics of mature neurons. J Alzheimers Dis 20(4):1069–1082. doi:10.3233/JAD-2010-091363
30. Cuende J, Moreno S, Bolanos JP, Almeida A (2008) Retinoic acid downregulates Rae1 leading to APC(Cdh1) activation and neuroblastoma SH-SY5Y differentiation. Oncogene 27(23):3339–3344. doi:10.1038/sj.onc.1210987
31. Arun P, Madhavarao CN, Moffett JR, Namboodiri AM (2008) Antipsychotic drugs increase N-acetylaspartate and N-acetylaspartylglutamate in SH-SY5Y human neuroblastoma cells. J Neurochem 106(4):1669–1680. doi:10.1111/j.1471-4159.2008.05524.x
32. Constantinescu R, Constantinescu AT, Reichmann H, Janetzky B (2007) Neuronal differentiation and long-term culture of the human neuroblastoma line SH-SY5Y. J Neural Transm Suppl 72:17–28
33. Guarnieri S, Pilla R, Morabito C, Sacchetti S, Mancinelli R, Fano G, Mariggio MA (2009) Extracellular guanosine and GTP promote expression of differentiation markers and induce S-phase cell-cycle arrest in human SH-SY5Y neuroblastoma cells. Int J Dev Neurosci 27(2):135–147. doi:10.1016/j.ijdevneu.2008.11.007
34. Gould J, Reeve HL, Vaughan PF, Peers C (1992) Nicotinic acetylcholine receptors in human neuroblastoma (SH-SY5Y) cells. Neurosci Lett 145(2):201–204
35. Lukas RJ, Norman SA, Lucero L (1993) Characterization of nicotinic acetylcholine receptors expressed by cells of the SH-SY5Y human neuroblastoma clonal line. Mol Cell Neurosci 4(1):1–12. doi:10.1006/mcne.1993.1001
36. Sokolova E, Matteoni C, Nistri A (2005) Desensitization of neuronal nicotinic receptors of human neuroblastoma SH-SY5Y cells during short or long exposure to nicotine. Br J Pharmacol 146(8):1087–1095. doi:10.1038/sj.bjp.0706426

Chapter 3

Cultured Cell Line Models of Neuronal Differentiation: NT2, PC12

Nune Darbinian

Abstract

The lack of a convenient, easily maintained and inexpensive in vitro human neuronal model to study neurodegenerative diseases, prompted us to develop a rapid, 1-h differentiated neuronal cell model based on human NT2 cells and C3 transferase. Here, we describe the rapid differentiation of human neuronal NT2 cells, and the differentiation, transduction and transfection of rat PC12 cells to obtain cells with the morphology of differentiated neurons that can express exogenous genes of interest at high level.

Key words Human neuronal cells, NT2, PC12, Differentiation, NGF, Rho kinase inhibitor, C3 transferase

1 Introduction

Developing of suitable and affordable neuronal models was always an important issue in neuroscience research. Although, several in vitro neuronal models have been proposed previously [1–3], all these techniques are time-consuming and require long-term (up to 4–6 weeks) incubations and expensive reagents. Differentiated human NT2 and rat PC12 cells may represent an excellent neuronal source for in vitro studies of neurodegenerative diseases, investigating pro-survival pathways and agents protecting human neuronal cells from cell death-mediating stresses.

The most ideal cell culture for analysis of the processes of neuronal differentiation would be one that can be differentiated rapidly so that it could be maintained easily and, at the same time, could be transfected with high efficiency to produce a stable or transient population of cells expressing exogenous gene products. These cells would promote extensive neuronal processes similar to that of primary neurons in culture. NTera (NT2), a human teratocarcinoma cells are capable of differentiating in response to retinoic acid, RA. To obtain pure neuronal cultures from RA-treated NT2 cells, further treatment with mitotic inhibitors should be

Shohreh Amini and Martyn K. White (eds.), *Neuronal Cell Culture: Methods and Protocols*, Methods in Molecular Biology, vol. 1078, DOI 10.1007/978-1-62703-640-5_3, © Springer Science+Business Media New York 2013

performed for at least 6 weeks [1]. These neurons (NT2-N cells) express all ubiquitous neuronal markers that can be identified in axons or dendrites using molecular and functional criteria. The aim of this study was to develop a neuronal culture system that can be easily maintained and can be rapidly differentiated. Thus, we further developed an in vitro neuronal model to obtain rapidly differentiated NT2 and PC12 cells.

2 Materials

1. NTera 2/cl.D1 (NT2), a human teratocarcinoma cell line. NT2 cells are obtained from American Type Culture Collection (ATCC, Manassas, VA).
2. PC12 rat pheochromocytoma cells (ATCC No. CRL-1721) cells are from American Type Culture Collection.
3. Retinoic acid, RA.
4. F-12 medium supplemented with 15 % horse serum and 2.5 % fetal bovine serum.
5. Dulbecco's Modified Eagle's Medium (DMEM), supplemented with 10 % fetal bovine serum (Life Technologies, Inc.) and antibiotics (100 U/ml penicillin and 10 μg/ml streptomycin).
6. Dulbecco's modified Eagle's medium high glucose (DMEM HG) supplemented with 10 % fetal bovine serum and penicillin/streptomycin.
7. DMEM for neurite regeneration with 10 % FBS, penicillin/streptomycin, supplemented with mitotic inhibitors (1 μM cytosine arabinoside and 10 μM uridine).
8. DMEM/F12 medium for PC12 cells containing 10 % fetal bovine serum (FBS) and 5 % horse serum (HS).
9. 37 °C humidified Incubator containing 7 % CO_2.
10. Collagen-coated 60 mm tissue culture dishes.
11. Serum-free F-12 medium.
12. NGF solution (100 ng/ml).
13. Cold phosphate-buffered saline (PBS).
14. pLECFP-C1 (Clontech, Mountain View, CA).
15. Commercially purchased synthetic peptide containing the arginine rich protein transduction domain (PTD) of HIV-1 Tat (amino-acids 47–57) fused to an N-terminal 6 histidine epitope tag.
16. Cover Glass (finest grade, Premium Hard, Glass/Clear/Pre-Cleared/Non-Corrosive/Non-Fogging, Thickness No 1 Size: 24 mm × 40 mm.

17. Chamber Slide Lab-Tec System Well Permanox Slide 4 chambers mounted on Permanox slide with cover.
18. Chamber Slide System and Chambered Coverglass Lab-TekII Chamber Slide.
19. 2 well Permanox slide.
20. Tissue Culture Dish Polystyrene 60 × 15 mm.
21. Tissue Culture Dish 100 × 20 mm Style.
22. Tryple Express enzyme (Invitrogen, Carlsbad, CA).
23. RhoA inhibitor, C3 transferase.

3 Methods

3.1 Cell Culture

3.1.1 Human Neurons

Ntera (NT2) human embryonic teratocarcinoma cells are maintained in Dulbecco's Modified Eagle's Medium (DMEM) at 37 °C in a humidified atmosphere containing 7 % CO_2. Cells are plated in T75 flasks at a density of one million cells, and 12× well dishes at a density of 3×10^4/well and cultured in 3 ml medium.

3.1.2 Postmitotic Terminally Differentiated Polarized Human Neurons

Postmitotic Terminally Differentiated Polarized Human Neurons are derived from NT2 according to published protocols [1–3] with some modifications. Cells are cultured in Dulbecco's modified Eagle's medium high glucose (DMEM HG). Cells are induced to differentiation upon incubation with RA 10 μM RA twice a week for 4 weeks. For neurite regeneration following RA treatment, cells are incubated in DMEM with mitotic inhibitors for 2–4 weeks. Then DMEM is replaced with Neurobasal medium. Differentiated neurons establish polarized neurites with the characteristics of axons and dendrites after 4 weeks of RA treatment and 2–4 weeks of incubation with mitotic inhibitors. Cells are maintained at 37 °C in a humidified incubator containing 7 % CO_2 (*see* **Note 1**).

3.1.3 PC12

PC12 rat pheochromocytoma cells are cultured on 60 mm poly-D-lysine-coated dishes at a density of 1×10^5 cells per 60 mm dishes in DMEM/F12 medium containing 10 % fetal bovine serum (FBS) and 5 % horse serum (HS). To induce neuronal differentiation, PC12 cells are plated on collagen IV-coated dishes and treated with 20 ng/ml nerve growth factor (NGF) in serum-free medium (SFM). Neuronal processes begin to form within first 24 h following the treatment and are preserved as differentiated neuronal cultures for 5 days (*see* **Note 2**).

3.2 Treatment

3.2.1 PC12 Cells

PC12 Cells are differentiated in the presence of NGF (20 ng/ml) for 5 days. The images of neurite-like processes were taken from series of photographs from selected microscopic fields and reflect an average length of neurite-like processes in a particular microscopic field.

3.2.2 NT2 Cell Treatment

NT2 cells are plated onto 12× well tissue culture plates at a density of 3×10^4 and cultured for 1 day. Cells are starved in serum free DMEM for 2 h prior to C3 transferase stimulation (1 μg/ml) for 1 h or 16 h. Each culture is also incubated with C3 transferase in the presence of serum [4].

3.3 Transfection

PC12 cells are transfected with pLEGFP-C1 plasmid (2 μg) using a high efficiency transfection reagent such a Lipofectamine 2000 (3 μl). Human neuronal cells are also transfected with pLEGFP-C1 expression plasmid (3 μg each) using a high efficiency transfection reagent (*see* **Note 3**).

3.4 Microscopy

Contrast and brightness were adjusted equally for all images using Adobe Photoshop version 5.5. Original magnification was 200×.

3.5 Antibodies

Expression of the GFP protein was examined by immunocytochemistry, using Living Colors full-length monoclonal antibody (BD Biosciences/Clontech). The presence of Protein Transduction Domain (PTD) from HIV-1 Tat protein fused to 6 Histidine amino acid sequence was examined by immunocytochemistry using anti-His(C-term)-tag Antibody (Invitrogen).

3.6 Protein Transduction and Fluorescence Staining

Approximately 1×10^5 PC12 cells are seeded in 60 mm dishes in medium containing fetal bovine and horse serum. After 8 h, medium is replaced with DMEM/F12 without serum for 16 h, after which the cells were treated with NGF for 48 h in 1 ml DMEM/F12 plus fetal bovine and horse serum. For staining, approximately 1,000 cells were seeded in poly L-lysine coated slide chambers and after 0, 2, and 24 h treatment with 25 μM of synthetic Tat PTD peptide tagged with six histidine amino acids, cells were fixed with cold acetone for 3 min and washed with PBS. Immunocytochemistry was performed by blocking with normal horse serum and incubating with anti-histidine antibody (Invitrogen) at 1:200 dilution for 16 h at room temperature. Cells were incubated with a fluoroscein labeled secondary antibody for 1 h. For nuclear counter staining, propidium iodide was included in the mounting medium. Protein Transduction Domain (PTD) was visualized in fluorescence green.

3.7 Expected Results and Conclusions

The lack of an easy maintained in vitro neuronal models to study human neurodegenerative diseases and signaling pathways that control neuronal differentiation, prompted us to develop a rapid, 1-h differentiated neuronal cell model based on NT2 human neuronal cells. Our aim was to develop an advanced in vitro model, generating sustainable cells with morphology of human, mature neurons during short period of time. First, we have developed a modified protocol to obtain a terminally differentiated NT2 cells that show the typical neuronal morphology with neuronal features,

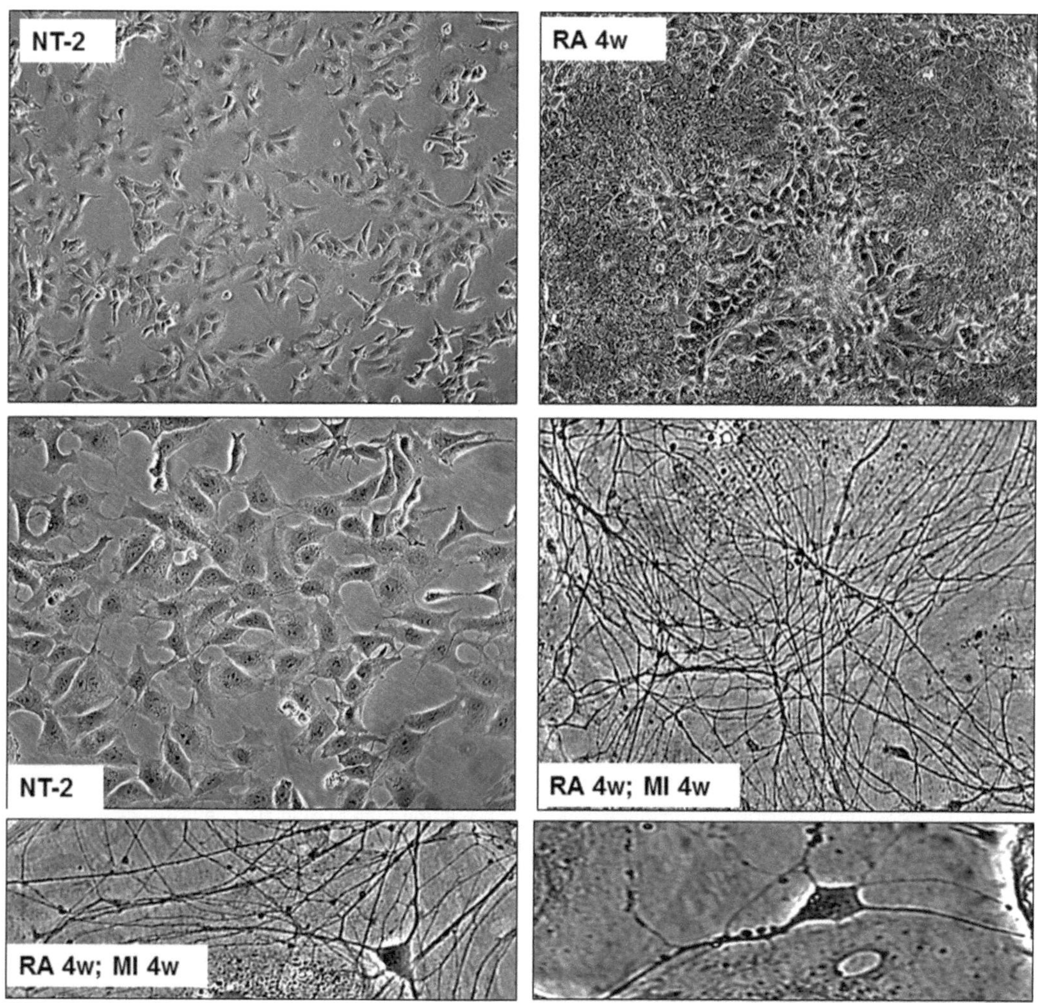

Fig. 1 Differentiation of neuronal NT2 cells. Phase images of undifferentiated and differentiated (8-week) human NT2 cells showing the morphologic changes during the terminal differentiation of NT2 cells. (*Top left panel*) Untreated NT2 cells (×1,000); (*second left panel*) untreated NT2 cells at higher magnification (×200); (*top right panel*) NT2 cells following 4-weeks of RA treatment; (*second right panel*) NT2 cells following replate and 4-weeks of treatment with mitotic inhibitors (note that most cells exhibit the morphology of neurons and extensive process outgrowth has occurred); (*bottom left panel*) terminally differentiated NT2 cells showing the typical neuronal morphology of these cells at higher magnification ×200 (note cell body and defined neuronal processes); (*bottom right panel*) single differentiated cell with neuronal features, the long process resembling an axon and the three major processes that resemble dendrites

the long process resembling an axon and the three major processes that resemble dendrites (Fig. 1). Inhibition of RhoA activation by a RhoA inhibitor, C3 transferase, also promotes massive long-lasting neurite outgrowth during short period of treatment of NT2 cells. Treatment of cells with C3 transferase produced extensive neurite outgrowth after as early as 1 h of incubation. Massive neurite outgrowth after 3 h of C-3 transferase treatment in

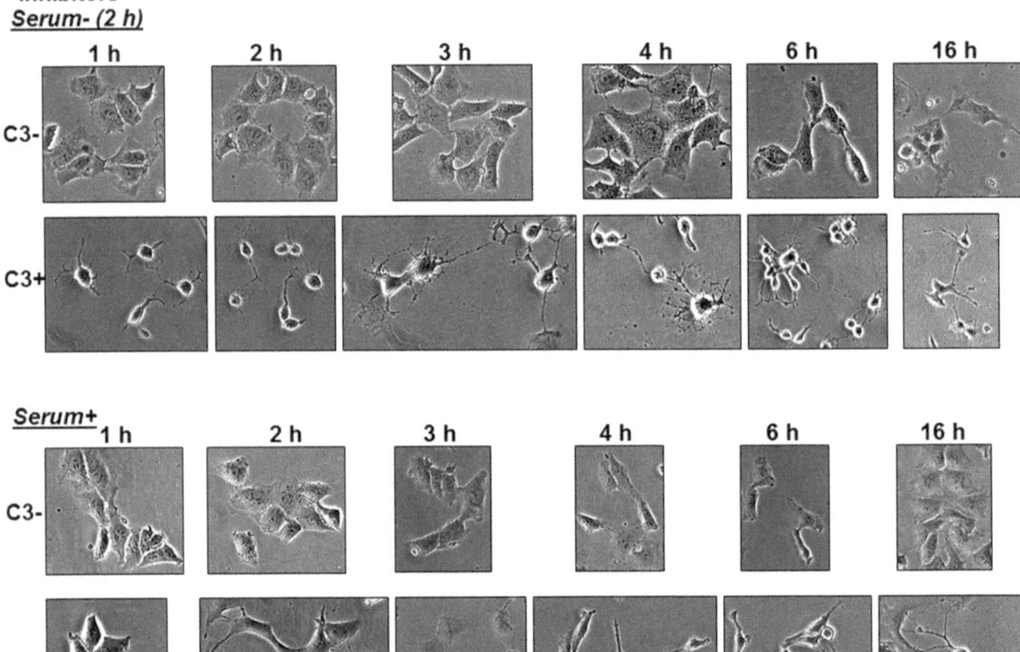

Fig. 2 Time-dependent effect of C3 transferase on neurite outgrowth in human NT2 neuronal cells. (*Top*) The effect of C3 transferase on the formation of neuronal processes in NT2 cells after 1 h incubation in serum-free medium. (*Bottom*) Neuronal outgrowth in NT2 cells, mediated by C3 transfeease in the presence of serum

GFP-expressing cells (Fig. 2). Interestingly, removal of C3 transferase does not lead to neurite retraction indicating that the C3 transferase in neuronal cells is critical for neuronal differentiation at early stages of treatment (Fig. 3). This model is useful to study signaling pathways resulting in neuronal cell injury. We also improved transfection protocols, specifically using Lipofectamine 2000, to efficiently transfect differentiated or undifferentiated neurons with various plasmids (Fig. 4). Immunocytochemical analysis of NT2 neuronal cells overexpressing GFP revealed cytoplasmic immunoreactivity for GFP and massive neuronal outgrowth (Fig. 5). Thus, it is possible transfect undifferentiated NT2 cells with expression plasmids allowing the introduction of expressed proteins into cells that can subsequently undergo induction to develop into stable, postmitotic, human neurons.

Another neuronal model, actively used by our group, is a PC12 rat neuronal cell line [5–7]. PC12 cells are differentiated upon NGF treatment (20 ng/ml) and can be used for rapid protein transduction assays. We have developed differentiation/protein transduction assay followed by immunocytochemistry and microscopy (Fig. 6). Results from PC12 differentiation and protein

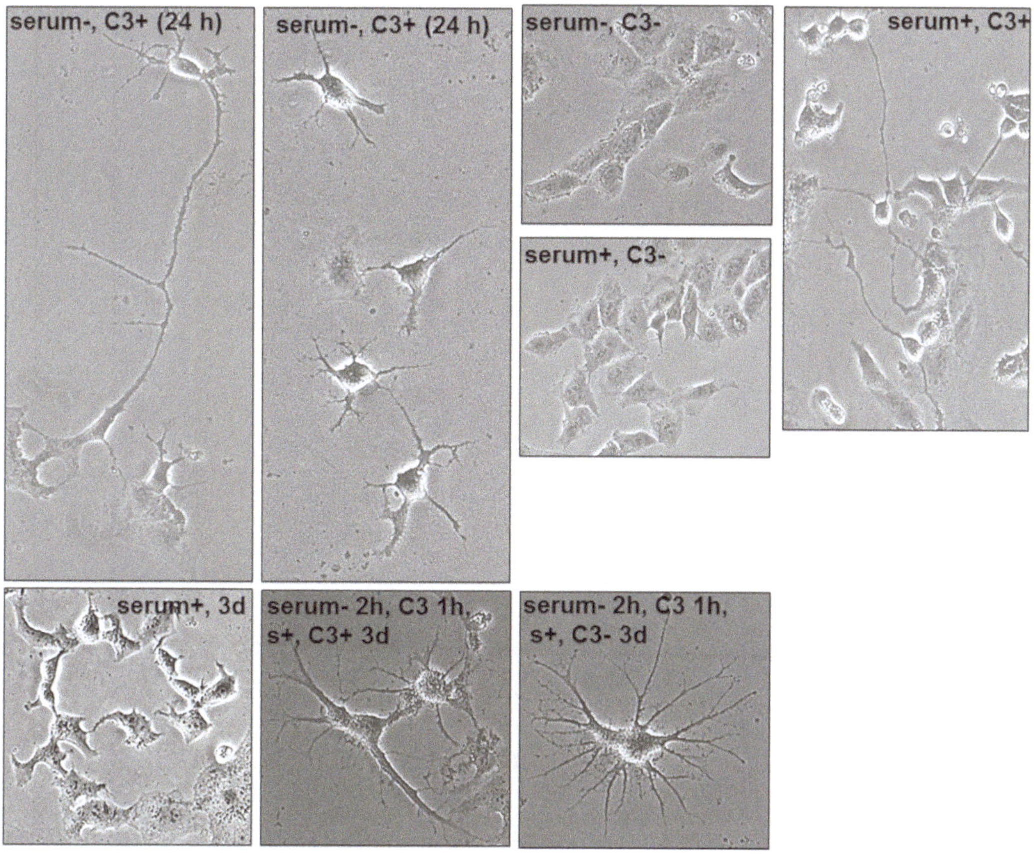

Fig. 3 Differentiation and rescue assays. Incubation of neuronal cells with C3 transferase longer than 1 h effects on. (*Top*) Neurite outgrowth in NT2 cells treated with the RhoA inhibitor, C3 transferase for 24 h. (*Bottom panels*) Removal of C3 transferase does not lead to neurite retraction, indicating that the C3 transferase in neuronal cells is critical for neuronal differentiation at early stages of treatment

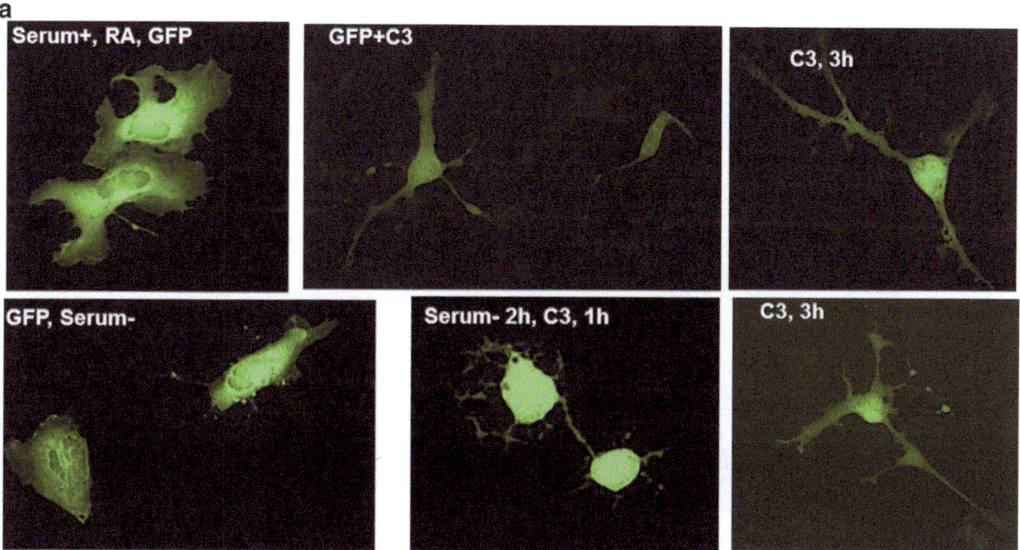

Fig. 4 NT2: Rapid differentiation, transfection, and neurite outgrowth. NT2 neuronal cells were transiently transfected with pLEGFP-C1 plasmid for 16 h and cells, after 2 h of serum starvation, were treated with C3

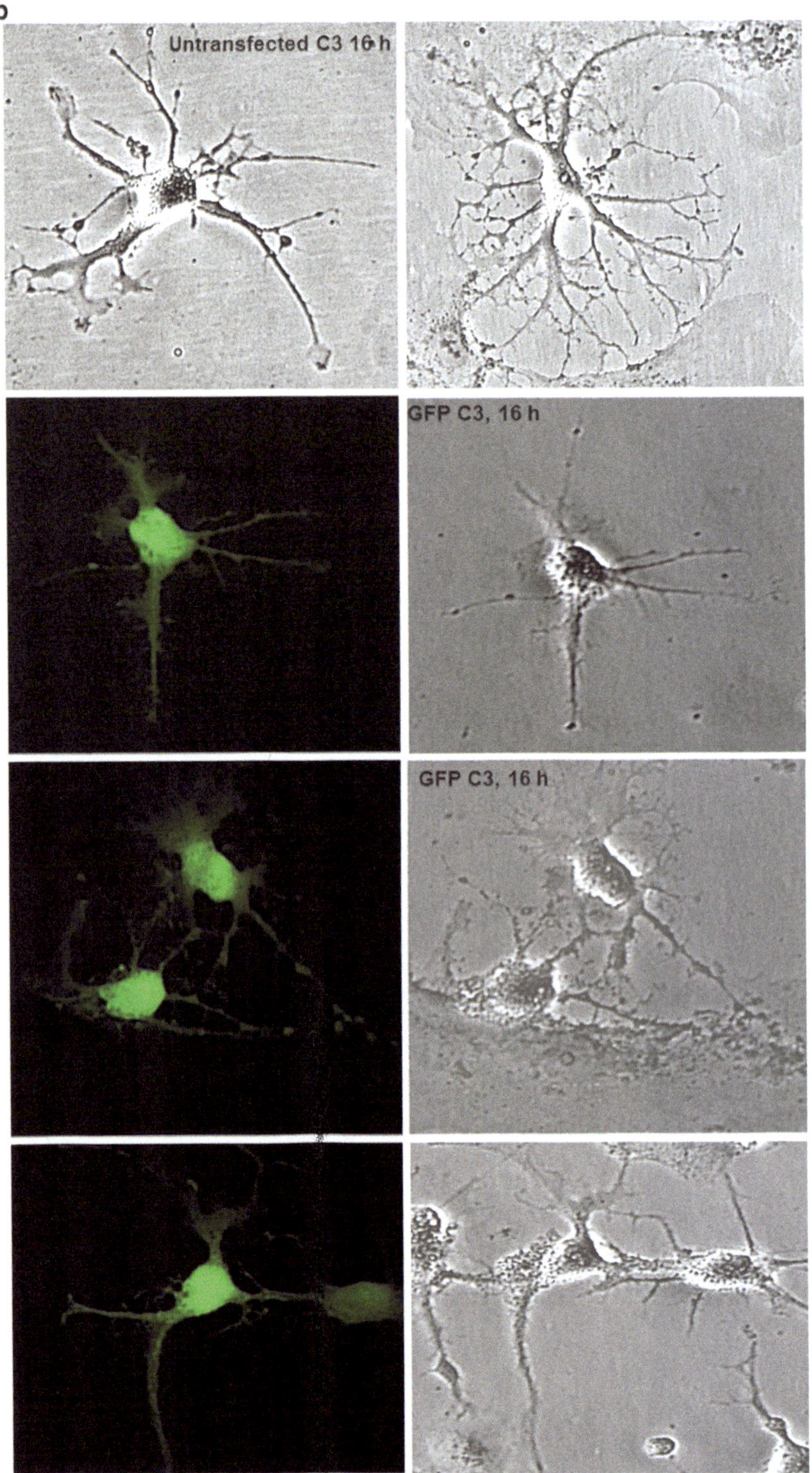

Fig. 4 (continued) transferase for 3 or 16 h to induce differentiation via RhoA inactivation. (**a**) Fluorescence images (magnification ×200) of neuronal cells transfected with GFP and incubated with C-3 transferase for 3 (**b**). Phase and fluorescence images of transfected and differentiated cells. Cells were kept in serum-free medium for 2 h and then incubated with C3 transferase for 16-h. C-3 transferase causes massive outgrowth in GFP-expressing cells

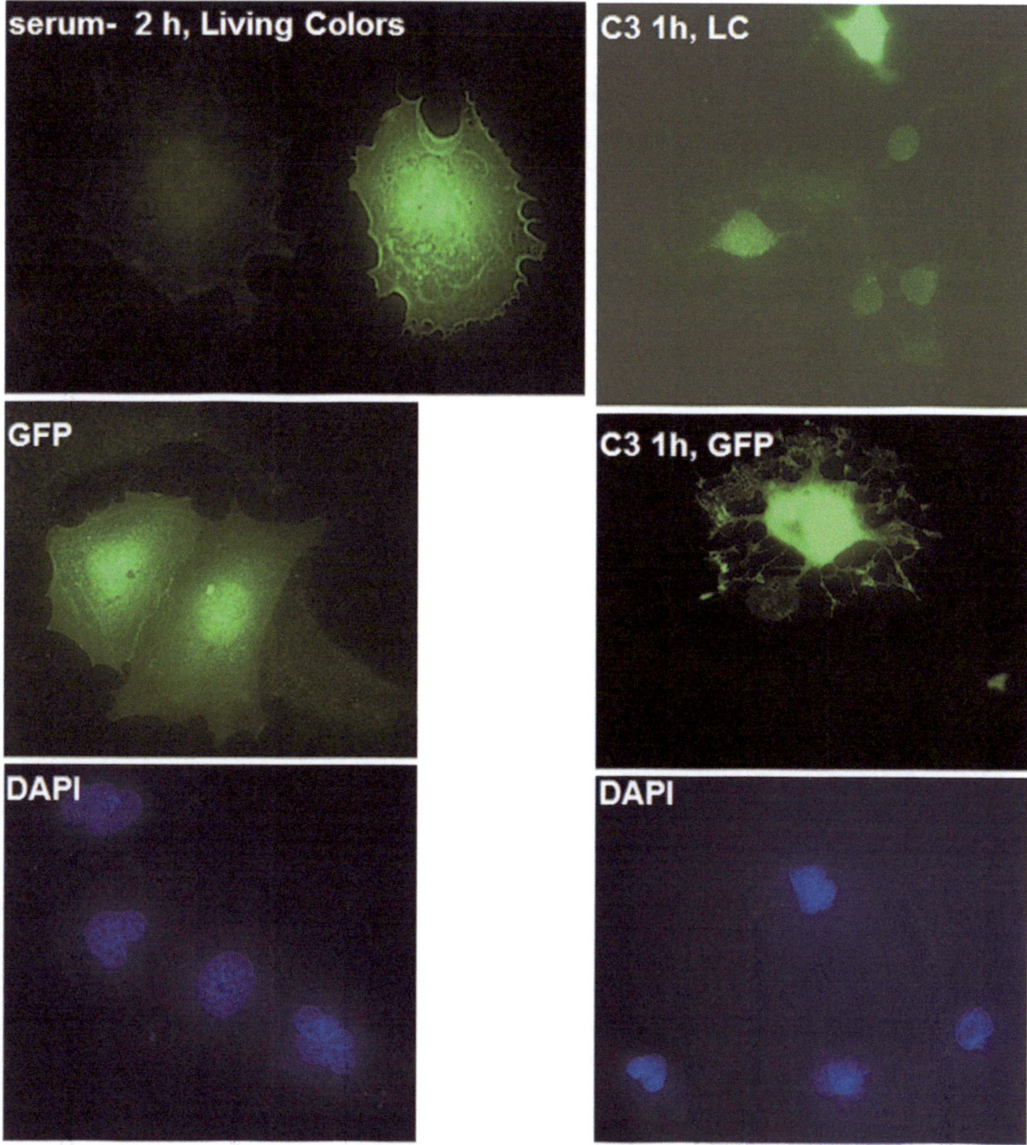

Fig. 5 NT2: Rapid differentiation, transfection, and immunostaining. Immunocytochemical analysis of NT2 neuronal cells overexpressing GFP and incubated with C3 transferase for 1 h. (*Top panels*) Immunostaining was performed with Living colors (for GFP protein) on cells with C3 transferase treatment (*right panels*) or without C3 incubation (*left panels*). Cytoplasmic immunoreactivity is seen for GFP. GFP-expressing cells demonstrate massive neuronal outgrowth (*right panels*). (*Bottom panels*) DAPI nuclear staining was performed in cells without or with C3 transferase

transduction experiments using a synthetic peptide containing the arginine rich Protein Transduction Domain (PTD) of HIV-1 Tat (aa 47–57) and a His Tag revealed cellular internalization and nuclear appearance of the PTD peptide after 2 h and its detection in nuclei up to 24 h after treatment. PC12 cells were also transfected with a plasmid expressing a gene of interest prior to differentiation (Fig. 7).

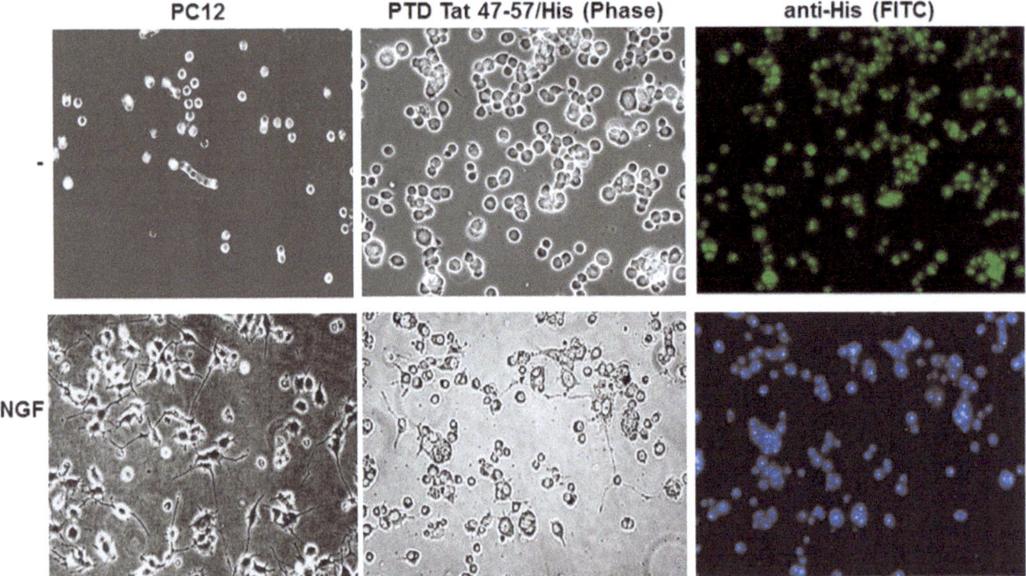

Fig. 6 PC12: Differentiation, protein transduction and detection. Undifferentiated (*top panels*) or differentiated PC12 cells (*bottom panels*) upon NGF treatment (20 ng/ml) were used for rapid protein transduction assays. We have developed differentiation/protein transduction assay followed by immunocytochemistry and microscopy. Results from PC12 differentiation and protein transduction experiments using the synthetic peptide representing the arginine rich Protein Transduction Domain (PTD) of HIV-1 Tat revealed cellular internalization and nuclear appearance of the PTD peptide visualized with anti-histidine antibody after 2 h and its detection in nuclei up to 24 h after treatment

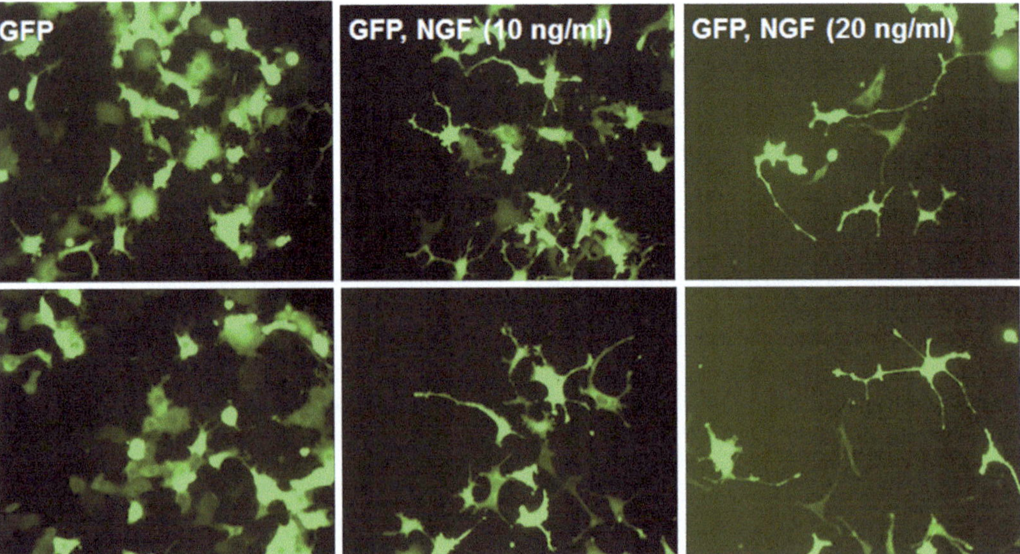

Fig. 7 PC12: differentiation and transient transfection. GFP-expressing PC12 cells were grown on 12× well plates (Falcon) for 1 day, then treated with increased concentrations of NGF (10–20 ng/ml). Results from transfection/differentiation experiments using two concentrations of NGF revealed robust cellular internalization of GFP and massive neuronal outgrowth after 48 h after treatment

Thus, NT2 cells and PC12 cells represent a unique model system for studies of human and rat neurons, and signaling pathways that control neuronal differentiation and represent novel models for the expression of diverse gene products in terminally differentiated polarized neurons, or robustly differentiated neuronal cells.

4 Notes

1. Make sure to add antibiotics to media for long incubation of NT2 cells before cells start to differentiate.
2. It is important to use fresh poly-D-lysine-coated dishes or collagen IV-coated dishes for efficient attachment and differentiation of PC12 cells.
3. The presence of antibiotics or serum in the media does not inhibit the transfection efficiency, if highly purified DNA used in transfection reactions.

References

1. Pleasure SJ, Page C, Lee VM (1992) Pure, postmitotic, polarized human neurons derived from NTera 2 cells provide a system for expressing exogenous proteins in terminally differentiated neurons. J Neurosci 12:1802–1815
2. Katoh M (2002) Regulation of WNT signaling molecules by retinoic acid during neuronal differentiation in NT2 cells: threshold model of WNT action (review). Int J Mol Med 10:683–687
3. Jain P, Cerone MA, Leblanc AC, Autexier C (2007) Telomerase and neuronal marker status of differentiated NT2 and SK-N-SH human neuronal cells and primary human neurons. J Neurosci Res 85:83–89
4. Mishra M, Del Valle L, Otte J et al (2012) Pur-alpha regulates RhoA developmental expression and downstream signaling. J Cell Physiol 228: 65–72
5. Darbinian N, Darbinyan A, Czernik M et al (2008) HIV-1 Tat inhibits NGF-induced Egr-1 transcriptional activity and consequent p35 expression in neural cells. J Cell Physiol 216: 128–134
6. Darbinian-Sarkissian N, Czernik M, Peruzzi F et al (2006) Dysregulation of NGF-signaling and Egr-1 expression by Tat in neuronal cell culture. J Cell Physiol 208:506–515
7. Peruzzi F, Gordon J, Darbinian N, Amini S (2002) Tat-induced deregulation of neuronal differentiation and survival by nerve growth factor pathway. J Neurovirol 8(Suppl 2): 91–96

Chapter 4

Murine Teratocarcinoma-Derived Neuronal Cultures

Prasun K. Datta

Abstract

This chapter describes the culture and propagation of murine embryonic stem cells, F9 and P19 and strategies for differentiation of these stem cells into neurons. Protocols focus on maintenance and propagation of these cells and routine procedures employed for differentiation into neuronal cells. Additional protocols are also described for obtaining enriched populations of mature neurons from P19 cells and differentiation of F9 cells into serotonergic or catecholaminergic neurons.

The protocols described herein can be employed for dissection of the pathways such as gliogenesis and neurogenesis that are involved in differentiation of pluripotent stem cells such as F9 and P19 into glial cells or terminally differentiated neurons.

Key words Embryonal carcinoma, Teratocarcinoma, Retinoic acid, P19, F9, Neurons, Differentiation

1 Introduction

Embryonal carcinoma (EC) cells are pluripotent stem cells of teratocarcinoma that can differentiate and give rise to all three primary germ layers: endoderm, mesoderm and ectoderm [1]. Currently, two mouse teratocarcinoma cell lines, F9 and P19 are widely used for differentiation into neuronal cells [2, 3]. Normal spontaneous differentiation of F9 and P19 cells is very low however the differentiation pathway can be induced by addition of retinoic acid (RA) or RA and dibutyryl cyclic AMP (dcAMP) into a variety of cell types including neuronal cells [4–9]. There are limitations in the use of RA as a differentiation agent for generation of neuronal cells since studies have shown that P19 cells yield various types of neurons as well as astrocytes, oligodendrocytes, and microglia after treatment with retinoic acid [10]. Thus, these stem cells are ideally suited for dissection of the differentiation pathway to a terminally differentiated neural phenotype. However, molecular studies arc hindered by the heterogeneity of differentiation.

Therefore, in recent years alternative strategies have been developed for both of these cell lines to induce differentiation into more mature neurons with significant enrichment of neuronal

Shohreh Amini and Martyn K. White (eds.), *Neuronal Cell Culture: Methods and Protocols*, Methods in Molecular Biology, vol. 1078, DOI 10.1007/978-1-62703-640-5_4, © Springer Science+Business Media New York 2013

population. It is worth mentioning that neurons derived from the P19 cells treated with RA express functional GABA receptors [11], and ionotropic glutamate receptors of both NMDA and AMPA/kainite types [12]. Furthermore, studies demonstrate that neurons derived from P19 cells mature in vivo and capable of displaying neuronal electrophysiological characteristics after 4 weeks of implantation into rat brains [13, 14]. P19 cells have also been used to understand the mechanism of mu-opioid receptor (MOR) upregulation during neuronal differentiation in P19 embryonal carcinoma cells and role of epigenetics in MOR upregulation [15, 16]

This chapter describes the techniques used for maintenance and expansion of both F9 and P19 cells (Basic Protocol A), and routinely used protocol for differentiation into neurons (Basic Protocol B), and then followed by recent protocols that involve modification of basic protocol B to differentiate into more mature enriched neuronal population (Specific protocols).

2 Materials

Prepare all solutions for use in tissue culture using tissue culture grade water. Use of ultrapure water (prepared by purifying deionized water to attain a sensitivity of 18 MΩ cm at 25 °C) is highly recommended for all other purposes. Tissue culture wares, flasks and dishes, cryovials. The described procedures are performed in a Class II biological laminar-flow hood.

Refrigerated Centrifuge, −20 °C and −70 °C Freezers and Tissue culture incubator.

2.1 Cell Lines

F9 cells (ATCC, VA, USA, CRL-1720).

P19 cells (ATCC, VA, USA, CRL-1825).

2.2 Reagents

1. DMEM.
2. α-MEM.
3. Fetal calf serum.
4. New born calf serum.
5. Penicillin/Streptomycin 100× stock solution (10,000 units of penicillin and 10,000 μg of streptomycin per ml).
6. Neurobasal-A medium.
7. N2 supplement (100×).
8. Dulbecco's PBS without calcium and magnesium.
9. All trans-retinoic acid.
10. Ethyl Alcohol.
11. Dibutyryl-cAMP.
12. Cyclohexane carboxylic acid.
13. DMSO.

2.3 Growth and Differentiation Media

1. F9 growth medium: DMEM, low glucose 90 %, fetal calf serum 10 %.
2. P19 growth medium: αMEM 90 %, Newborn calf serum 7.5 %, fetal calf serum 2.5 %.
3. F9 differentiation medium: DMEM, 95 %, fetal calf serum 5 %.
4. P19 differentiation medium: α-MEM, fetal calf serum 5 %.
5. Freeze media: Growth media containing 10 % (v/v) tissue culture grade dimethyl sulfoxide (DMSO).

2.4 Preparation of Retinoic Acid

Dissolve all trans-retinoic acid (RA) in ethanol to a stock solution of 3 mg/ml (0.01 M) (*see* **Note 4**).

2.5 Preparation of Dibutyryl-cAMP (db-cAMP)

db-cAMP is prepared as 10^{-2} M (10×) or 2×10^{-2} M (20×) stock directly in tissue culture media. Filter the dissolved solution using a 0.2 μm filter before use and is recommended to be used immediately.

3 Methods

3.1 Basic Protocol A

3.1.1 Maintenance and Propagation of F9 Cells in Culture

1. Remove and discard culture medium and wash cells with Dulbecco-PBS to remove residual serum from culture media (*see* **Notes 1** and **2**).
2. Add 1–2 ml of 0.25 % Trypsin/EDTA solution and observe cells under an inverted microscope until cell layer is dispersed (usually within 5–15 min) (*see* **Note 3**).
3. Add 8–10 ml of complete growth medium to inactivate trypsin and aspirate cells by gently pipetting.
4. Transfer cell suspension into 15 ml conical centrifuge tube and centrifuge cells for 5 min at 500 × *g*.
5. Discard supernatant and resuspend cells in 10 ml of F9 growth medium, count the number of cells using a Neubauer hemocytometer.
6. Plate cells at a minimal density of 6×10^5 cells/cm^2 in tissue culture dish or tissue culture flask.
7. Incubate cultures in a humidified tissue culture incubator at 37 °C with 5 % CO_2.

3.1.2 Maintenance and Propagation of P19 Cells in Culture

1. Remove and discard culture medium and wash undifferentiated cells (Fig. 1a) with Dulbecco-PBS to remove residual serum from culture media (*see* **Notes 1** and **2**).
2. Add 1–2 ml of 0.25 % Trypsin/EDTA solution and observe cells under an inverted microscope until cell layer is dispersed (usually within 5–15 min) (*see* **Note 3**).

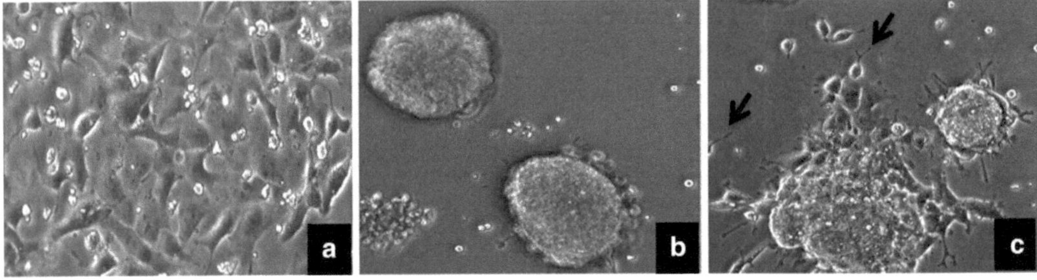

Fig. 1 Neural differentiation of P19 murine embryonic carcinoma cells. (**a**) Undifferentiated P19 cells. (**b**) Embryonic body (EB) formation, P19 cells treated with RA for 2 days and plated on non-adherent petri dishes. (**c**) EB bodies plated on day 3 on adherent tissue culture dish, note migration of cells from EB and formation of neurites in individual cells (*arrows*)

3. Add 8–10 ml of complete growth medium to inactivate trypsin and aspirate cells by gently pipetting.
4. Transfer cell suspension into 15 ml conical centrifuge tube and centrifuge cells for 5 min at 500×*g*.
5. Discard supernatant and resuspend cells in 10 ml of P19 growth medium, count the number of cells using a Neubauer hemocytometer.
6. Plate cells at a minimal density of 6×10^5 cells/cm^2 in tissue culture dish or flask.
7. Incubate cultures in a humidified tissue culture incubator at 37 °C with 5 % CO_2.

3.1.3 Freezing Cells

1. Remove and discard culture medium and wash cells with Dulbecco-PBS to remove residual serum from culture media.
2. Add 1–2 ml of 0.25 % Trypsin/EDTA solution and observe cells under an inverted microscope until cell layer is dispersed (usually within 5–15 min).
3. Add 8–10 ml of complete growth medium to inactivate trypsin and aspirate cells by gently pipetting.
4. Transfer cell suspension into 15 ml conical centrifuge tube and centrifuge cells for 5 min at 500×*g*.
5. Discard supernatant and resuspend cells in 5 ml of P19 or F9 growth medium, count the number of cells using a Neubauer hemocytometer.
6. Pellet cells by centrifugation for 5 min at 500×*g*.
7. Resuspend cells at 1–2×10^6 cells/ml in freezing medium, mix gently, and transfer 1 ml of aliquots into labeled cryovials.
8. Store cells overnight at −70 °C, then transfer vials to liquid nitrogen cryotank for long-term storage.

3.2 Basic Protocol B

This protocol is widely used to differentiate F9 cells into neurons and is based on a published protocol [17].

3.2.1 Induction of Neuronal Differentiation of F9 Cells

1. Remove culture medium from flask containing F9 cells (75–80 % confluency), wash cells with PBS (*see* **Notes 1** and **2**).
2. Add 1–2 ml of 0.25 % Trypsin/EDTA solution and observe cells under an inverted microscope until cell layer is dispersed (usually within 5–15 min) (*see* **Note 3**).
3. Add 8–10 ml of complete growth medium to inactivate trypsin and aspirate cells by gently pipetting.
4. Transfer cell suspension into 15 ml conical centrifuge tube and centrifuge cells for 5 min at $500\times g$.
5. Discard supernatant and resuspend cells in 10 ml of P19 growth medium, count the number of cells using a Neubauer hemocytometer.
6. Plate 1×10^6 cells in 10 ml of F9 differentiation media in tissue culture dish or flask and induce differentiation by adding RA and dcAMP solutions to achieve a final concentration of 10^{-7} M and 10^{-3} M, respectively (*see* **Note 5**).
7. Rock the suspension gently to ensure even mixing of cells and RA. Leave the dish/flask undisturbed in a humidified tissue culture incubator at 37 °C with 5 % CO_2 for 10 days.
8. Change media in the dish/flask every 2–3 days. Continue adding RA and dcAMP after each media change as described in **step 6**.

3.2.2 Induction of Neuronal Differentiation of P19 Cells

This protocol is used for differentiation of P19 cells into mature neurons and involves generation of cell aggregates popularly known as embryonic bodies (EB) using cells plated on non-adherent culture ware in presence of RA, the aggregates are then plated in presence of RA in regular tissue culture ware.

1. Remove culture medium from flask containing P19 cells (75–80 % confluency), wash cells with PBS (*see* **Notes 1** and **2**).
2. Add 1–2 ml of 0.25 % Trypsin/EDTA solution and observe cells under an inverted microscope until cell layer is dispersed (usually within 5–15 min) (*see* **Note 3**).
3. Add 8–10 ml of complete growth medium to inactivate trypsin and aspirate cells by gently pipetting.
4. Transfer cell suspension into 15 ml conical centrifuge tube and centrifuge cells for 5 min at $500\times g$.
5. Discard supernatant and resuspend cells in 10 ml of P19 growth medium, count the number of cells using a Neubauer hemocytometer.

6. Plate 1×10^6 cells in 10 ml of P19 differentiation media in 100 mm × 10 mm bacteria grade petri dish and induce differentiation by adding 1 μl of stock RA solution to 10 ml medium to attain a final concentration of 1 μM RA.
7. Rock the suspension gently to ensure even mixing of cells and RA. Leave the petri dishes undisturbed in a humidified tissue culture incubator at 37 °C with 5 % CO_2 for 2 days for formation of EB (Fig. 1b).
8. On day 2 harvest the EB by transferring them to a 50 ml conical tube and allow the aggregates or EB to settle down at room temperature in the tissue culture hood for 15–20 min.
9. Remove the media and then add 10 ml of P19 growth media and transfer cell suspension to regular tissue culture flask or dish and incubate in a humidified tissue culture incubator at 37 °C with 5 % CO_2 for 2 days.
10. On day 2 remove the media from culture ware and add 10 ml of P19 differentiation media and incubate in a humidified tissue culture incubator at 37 °C with 5 % CO_2 for 7–8 days, replace half of the media every 2 days to allow maturation of neurons (Fig. 1c) (*see* **Note 8**).

3.3 Specific Protocols

This section describes recent protocols for differentiation of F9 and P19 cells into more mature neurons with significant enrichment of neuronal population.

3.3.1 Protocol for Generation of Serotonergic or Catecholaminergic Neurons from F9 Cells

This protocol described by Mouillet-Richard et al. [18] utilizes a F9 derived cell line that was generated by introduction of the early genes of SV40 under the control of the adenovirus E1A promoter [19] known as 1C11 clone to differentiate into serotonergic or catecholaminergic neurons. The advantage of using 1C11 cells is that one can attain pure population of neuronal cells that convert within 4 days of induction into serotonergic cells and are able to metabolize, store, and take up serotonin (5-HT)[1] [20] and express $5\text{-HT}_{1B/D}$, 5-HT_{2A}, and 5-HT_{2B} receptors [21]. Similarly, with appropriate induction the 1C11 cells within 12 days display a complete catecholaminergic phenotype, coincident with the induction of a functional norepinephrine (NE) uptake (*see* **Note 9**).

1. Remove culture medium from flask containing 1C11 cells (75–80 % confluency), wash cells with D-PBS.
2. Add 1–2 ml of 0.25 % Trypsin/EDTA solution and observe cells under an inverted microscope until cell layer is dispersed (usually within 5–15 min).
3. Add 8–10 ml of complete F9 growth medium to inactivate trypsin and aspirate cells by gently pipetting.
4. Transfer cell suspension into 15 ml conical centrifuge tube and centrifuge cells for 5 min at 500 × *g*.

5. Discard supernatant and resuspend cells in 10 ml of F9 growth medium, count the number of cells using a Neubauer hemocytometer.
6. Plate 4×10^4 cells/cm^2 in F9 growth media containing 10 % FBS (*see* **Note 7**) and supplemented with 1 mM dbcAMP and 0.05 % cyclohexane carboxylic acid (CCA) in tissue culture dish or flask to differentiate into serotonergic neurons.
7. Leave the dish/flask in a humidified tissue culture incubator at 37 °C with 5 % CO_2 for 4 days.
8. Change media in the dish/flask every 2 days. Continue adding 1 mM dbcAMP and 0.05 % CCA after each media change as described in **step 6**.
9. Plate 4×10^4 cells/cm^2 in DMEM supplemented with 5-HT depleted 10 % FBS and 1 mM dbcAMP, 0.05 % CCA and 2 % DMSO in tissue culture dish or flask to differentiate into catecholaminergic neurons.
10. Leave the dish/flask in a humidified tissue culture incubator at 37 °C with 5 % CO_2 for 12 days.
11. Change media in the dish/flask every 2 days with DMEM supplemented with 5-HT depleted 10 % FBS and 1 mM dbcAMP, 0.05 % CCA and 2 % DMSO.

3.3.2 Protocol for Generation of High-Yield Enriched Neuronal Cultures from P19 Cells

The cell culture protocol described in the following section is adapted from the protocol described by Monzo et al. [22] and has been demonstrated to generate a large number of fully mature neurons at both morphological and electrophysiological levels.

1. Remove culture medium from flask containing P19 cells (75–80 % confluency), wash cells with D-PBS.
2. Add 1–2 ml of 0.25 % Trypsin/EDTA solution and observe cells under an inverted microscope until cell layer is dispersed (usually within 5–15 min).
3. Add 8–10 ml of complete growth medium to inactivate trypsin and aspirate cells by gently pipetting.
4. Transfer cell suspension into 15 ml conical centrifuge tube and centrifuge cells for 5 min at $500 \times g$.
5. Discard supernatant and resuspend cells in 10 ml of P19 growth medium, count the number of cells using a Neubauer hemocytometer.
6. Plate 6×10^4 cells/cm^2 in P19 differentiation media supplemented with 1 μM RA in tissue culture dish or flask.
7. Incubate cells in a humidified tissue culture incubator at 37 °C with 5 % CO_2 for 4 days.
8. Replace media with P19 differentiation media supplemented with 1 μM RA after 2 days.

9. After 4 days in culture wash the cells two to three times with culture medium and trypsinize the cells as described in **steps 2–4**.
10. Resuspended cell pellet obtained in **step 5** in P19 growth medium, recentrifuge cells, and wash cell pellet twice with Neurobasal-A medium without supplements.
11. Resuspend cells in NBA-medium supplemented with N2 supplement and 2 mM GlutaMAX.
12. Plate 9×10^4 cells/cm^2 on Matrigel coated plates (*see* Subheading 4, *see* **Note 6**) for preparing plates, if you wish to coat your own plates.
13. Incubate plated cells in a humidified tissue culture incubator at 37 °C with 5 % CO_2 for 20 days with medium changes on day 2 and 4 with NBA-medium supplemented with N2 and GlutaMAX.
14. On day 5, replace culture medium with NBA-medium supplemented with 2 mM GlutaMAX, 1×B27, 8 μM Cytosine-D-arabinofuranoside (AraC) and 8 nM 2′-deoxycytidine (2dCTD; Sigma) for the following 5 days. Replace media every 2 days.
15. On day 10 replace cell culture medium with NBA-medium supplemented only with 2 mM GlutaMAX and 1× B27 until day 20. Change cell culture medium every 2 or 3 days.

4 Notes

1. Culture for both cell lines must be maintained in the exponential growth phase. F9 and P19 cells grow rapidly with a generation time of 10–12 and 12–14 h, respectively. Since these cells do not regulate their multiplication in a density-dependent manner it is highly recommended that cells are subcultured at a ratio of 1:10 with renewal of media every 2–3 days.
2. Do not allow cells to be too confluent before the initiation of differentiation.
3. To avoid clumping do not agitate the cells by hitting or shaking the flask while waiting for the cells to detach. Cells that are difficult to detach may be placed at 37 °C to facilitate dispersal.
4. Sterile filter high grade ethanol used for dissolving RA. RA is light sensitive to light, wrap tubes containing stock solution with aluminum foil and store it at −80 °C for long-term storage. Stock solution is stable for 1 month if stored at −20 °C. Vortex the stock solution of RA to ensure that any precipitate that may have formed is redissolved. An alternative approach to redissolve RA faster is to leave the stock solution at room temperature for 15 min.

5. An alternative approach of increasing the intracellular concentration of cAMP for differentiation of F9 cells is to use a combination of cholera toxin at a final concentration of 10^{-10} M and the cyclic phosphodiesterase inhibitor, 3-isobutyl-L-methylxanthine (IBMX) at a final concentration of 10^{-4} M [23].
6. Dilute Matrigel stock solution in the ratio 1:45 in NBA-medium and dispense a volume of 0.1 ml/cm^2 of desired plate/flask. Leave plates/flask at room temperature for about 1 h, aspirate the coating solution and leave plates/flask to dry at room temperature for about 15 min. Store coated plates sealed at 4 °C.
7. It is important to deplete 5-HT from medium used for differentiation of P19 cells into serotonergic neurons by dialysis of culture media supplemented with 10 % FCS to less than 1 nM [18].
8. If desired, the status of neuronal differentiation can be assessed by immunohistochemistry using antibodies for neuronal markers such as β-tubulin III, enolase, neurofilament 200, and MAP2 [18].
9. Differentiation of 1C11 cells into serotonergic or catecholaminergic neurons can be also assessed by immunohistochemistry using antibodies for 5-HT or norepinephrine, respectively [21].

Acknowledgments

I am grateful to Dr. Ron Reichel for introducing me to the field of embryonic stem cells and differentiation during the early years of my career. Studies in the author's laboratory are funded by a grant from NIH/NIDA.

References

1. Martin GR (1980) Teratocarcinomas and mammalian embryogenesis. Science 209: 768–776
2. Bernstine EG, Hooper ML, Grandchamp S, Ephrussi B (1973) Alkaline phosphatase activity in mouse teratoma. Proc Natl Acad Sci U S A 790:3899–3903
3. McBurney MW, Rogers BJ (1982) Isolation of male embryonal carcinoma cells and their chromosome replication patterns. Dev Biol 89:503–508
4. Martin CA, Ziegler LM, Napoli JL (1990) Retinoic acid, dibutyryl-cAMP, and differentiation affect the expression of retinoic acid receptors in F9 cells. Proc Natl Acad Sci U S A 87:4804–4808
5. Jones-Villeneuve EM, McBurney MW, Rogers KA, Kalnins VI (1982) Retinoic acid induces embryonal carcinoma cells to differentiate into neurons and glial cells. J Cell Biol 94: 253–262
6. Kuff EL, Fewell JW (1980) Induction of neural-like cells and acetylcholinesterase activity in cultures of F9 teratocarcinoma treated with retinoic acid and dibutyryl cyclic adenosine monophosphate. Dev Biol 77:103–115
7. Strickland S, Sawey MJ (1980) Studies on the effect of retinoids on the differentiation of teratocarcinoma stem cells in vitro and in vivo. Dev Biol 78:76–85
8. Liesi P, Rechardt L, Wartiovaara J (1983) Nerve growth factor induces adrenergic neuronal

differentiation in F9 teratocarcinoma cells. Nature 306:265–267
9. Datta PK, Budhiraja S, Reichel RR, Jacob ST (1997) Regulation of ribosomal RNA gene transcription during retinoic acid-induced differentiation of mouse teratocarcinoma cells. Exp Cell Res 231:198–205
10. St-Arnaud R, Craig J, McBurney MW, Papkoff J (1989) The int-1 proto-oncogene is transcriptionally activated during neuroectodermal differentiation of P19 mouse embryonal carcinoma cells. Oncogene 4:1077–1080
11. Macpherson PA, Jones S, Pawson PA, Marshall KC, McBurney MW (1997) P19 cells differentiate into glutamatergic and glutamate-responsive neurons in vitro. Neuroscience 80: 487–499
12. Magnuson DS, Morassutti DJ, McBurney MW, Marshall KC (1995) Neurons derived from P19 embryonal carcinoma cells develop responses to excitatory and inhibitory neurotransmitters. Brain Res Dev Brain Res 90: 141–150
13. Morassutti DJ, Staines WA, Magnuson DS, Marshall KC, McBurney MW (1994) Murine embryonal carcinoma-derived neurons survive and mature following transplantation into adult rat striatum. Neuroscience 58:753–763
14. Magnuson DS, Morassutti DJ, Staines WA, McBurney MW, Marshall KC (1995) In vivo electrophysiological maturation of neurons derived from a multipotent precursor (embryonal carcinoma) cell line. Brain Res Dev Brain Res 84:130–141
15. Hwang CK, Kim CS, Kim do K, Law PY, Wei LN, Loh HH (2010) Up-regulation of the mu-opioid receptor gene is mediated through chromatin remodeling and transcriptional factors in differentiated neuronal cells. Mol Pharmacol 78:58–68
16. Hwang CK, Song KY, Kim CS, Choi HS, Guo XH, Law PY, Wei LN, Loh HH (2009) Epigenetic programming of mu-opioid receptor gene in mouse brain is regulated by MeCP2 and Brg1 chromatin remodelling factor. J Cell Mol Med 13:3591–3615
17. Darrow AL, Rickles RJ, Strickland S (1990) Maintenance and use of F9 teratocarcinoma cells. Methods Enzymol 190:110–117
18. Mouillet-Richard S, Mutel V, Loric S, Tournois C, Launay JM, Kellermann O (2000) Regulation by neurotransmitter receptors of serotonergic or catecholaminergic neuronal cell differentiation. J Biol Chem 275: 9186–9192
19. Kellermann O, Kelly F (1986) Immortalization of early embryonic cell derivatives after the transfer of the early region of simian virus 40 into F9 teratocarcinoma cells. Differentiation 32:74–81
20. Buc-Caron MH, Launay JM, Lamblin D, Kellermann O (1990) Serotonin uptake, storage, and synthesis in an immortalized committed cell line derived from mouse teratocarcinoma. Proc Natl Acad Sci U S A 87: 1922–1926
21. Kellermann O, Loric S, Maroteaux L, Launay JM (1996) Sequential onset of three 5-HT receptors during the 5-hydroxytryptaminergic differentiation of the murine 1C11 cell line. Br J Pharmacol 118:1161–1170
22. Monzo HJ, Park TI, Montgomery JM, Faull RL, Dragunow M, Curtis MA (2012) A method for generating high-yield enriched neuronal cultures from P19 embryonal carcinoma cells. J Neurosci Methods 204:87–103
23. Goldstein B, Rogelj S, Siegel S, Farmer SR, Niles RM (1990) Cyclic adenosine monophosphate-mediated induction of F9 teratocarcinoma differentiation in the absence of retinoic acid. J Cell Physiol 143:205–212

Chapter 5

Isolation and Propagation of Primary Human and Rodent Embryonic Neural Progenitor Cells and Cortical Neurons

Armine Darbinyan, Rafal Kaminski, Martyn K. White, Nune Darbinian, and Kamel Khalili

Abstract

The research on human neural progenitor cells holds great potential for the understanding the molecular programs that control differentiation of cells of glial and neuronal lineages and pathogenetic mechanisms of neurological diseases. Stem cell technologies provide also opportunities for pharmaceutical industry to develop new approaches for regenerative medicine. Here we describe the protocol for isolation and maintenance of neural progenitor cells and cortical neurons using human fetal brain tissue. This protocol can be successfully adapted for preparation of rodent neural and oligodendrocyte progenitor cells. While several methods for isolation of neural and oligodendrocyte progenitors from rodent brain tissue have been described, including techniques which use gene transfer and magnetic resonance beads, few methods are focused on derivation of human oligodendrocyte progenitor cells. Development of human culture provides the most physiologically relevant system for investigation of mechanisms which regulate function of oligodendrocyte, specifically of human origin.

Key words Human neural progenitor cells, Neurosphere, Oligodendrocyte progenitor cells, Neuron, Oligodendrocyte

1 Introduction

We developed protocols for preparation of fetal cortical neurons, human embryonic neural progenitor cells (hNPC) and their differentiation into oligodendrocyte lineage based on studies published studies [1–4]. Cultures of hNPC are prepared from human fetal brain tissue (embryonic age 16 weeks). Rodent NPC are isolated from the brain of timed-pregnant, embryonic day 16 or 17 (ED16 or ED17) Sprague–Dawley rat. It is important to note that while hNPC can be propagated several times and can be cryopreserved, primary neurons should be utilized without splitting and cryopreservation. Oligodendrocyte lineage progression is achieved by adjusting cell culture medium with defined growth factors required at particular developmental stage of cells [5]. In our protocol differentiation

Shohreh Amini and Martyn K. White (eds.), *Neuronal Cell Culture: Methods and Protocols*, Methods in Molecular Biology, vol. 1078, DOI 10.1007/978-1-62703-640-5_5, © Springer Science+Business Media New York 2013

of cells is achieved without gene-transfer manipulations providing a platform for studying the cellular behavior in "natural" conditions. This system is also can be used for investigation of more complex neuroglial interactions in co-culture experiments [6].

2 Materials

Large number of hNPC and primary cortical neurons can be obtained from one fetal brain (approximately 10.0–11.0 g in average). Five to eight rat embryos produce acceptable number of NPC. The following protocol describes methods for isolation of NSC from human fetal brain. Carry out all procedures in sterile conditions at room temperature unless otherwise specified.

2.1 Brain Tissue

Human fetal brain is obtained from Advanced Bioscience Resources (ABR), Inc., Alameda, CA 94501 USA in accordance with NIH guidclincs.

One timed-pregnant, embryonic day 16 or 17 (ED16 or ED17) Sprague–Dawley rat (Charles River Laboratories).

Protocols using human tissue and animals must be approved by an Institutional Animal Care and Use Committee (IACUC).

2.2 Reagents

Tryple™ Express (Gibco, cat# 12605).

DNase I (Sigma, AMPD1-1KT, St. Louis, MO).

Phosphate-buffered saline (1× PBS; Sigma, cat # P-5368).

anti-PSA-NCAM antibody (Developmental Studies Hybridoma Bank, University of Iowa, DSHB, 5A5).

Poly-D-lysine (Sigma, St. Louis, MO).

Cytosine arabinoside (Ara-C, C1768 Sigma, St. Louis, MO).

Immunopanning cocktail (50 mM Tris–Cl, pH 9.5/1 % BSA (w/v)/50 μg/ml anti-PSA-NCAM antibody).

Roche Applied Sciences Sybr Green mix (cat # 04707516001).

Neural stem cell medium (NSCM):

Neurobasal medium (Invitrogen, # 10888-022).

N2 supplement (Invitrogen, # 17502048).

B27 supplement without vitamin A (Invitrogen, #12587-010).

GlutaMAX (Invitrogen, 100×, cat # 35050061).

10 ng/ml Recombinant Human Fibroblast Growth Factor Basic (bFGF, Invitrogen, PHG0264).

10 ng/ml Recombinant Human Epidermal Growth Factor (EGF, Invitrogen, PHG0311).

10 ng/ml Recombinant Human Insulin Like Growth Factor (IGF-1, Invitrogen, PHG0078), Penicillin/streptomycin (Sigma, 100×, cat # P4333).

Gentamicin solution, 50 μg/ml final (Sigma, cat # G1272).

Amphotericin B, 2.5 mg/l final (Sigma, cat # A2942) (*see* **Note 1**).

Oligodendrocyte medium (OM): 1 part of NSCM and 1 part of glial medium with growth factors (GM+).

Glia medium with growth factors (GM+):

DMEM/F12 (1:1) high glucose (Gibco, cat # 11330).

N2 supplement (Invitrogen, # 17502048).

GlutaMAX (Invitrogen, 100×, cat # 35050061).

20 ng/ml bFGF (Invitrogen, PHG0264).

20 ng/ml EGF (Invitrogen, PHG0311).

20 ng/ml Recombinant human Platelet Derived Growth Factor-AA (PDGF-AA, Invitrogen, cat # PHG0035).

Penicillin/streptomycin (Sigma, 100×, cat # P4333).

Gentamicin solution, 50 μg/ml final (Sigma, cat # G1272).

Amphotericin B, 2.5 mg/l final (Sigma, cat # A2942).

Neuronal medium

Neurobasal medium (Invitrogen, # 10888-022).

B27 supplement with vitamin A (Invitrogen, cat # 0080085SA).

GlutaMAX (Invitrogen, 100×, cat # 35050061).

5 ng/ml Human Recombinant Brain-Derived Neurotrophic factor (BDNF, Invitrogen, cat # 10908010).

Penicillin/streptomycin (Sigma, 100×, cat # P4333).

Gentamicin solution, 50 μg/ml final (Sigma, cat # G1272).

Amphotericin B, 2.5 mg/l final (Sigma, cat # A2942).

2.3 Dissection Instruments, Plasticware, and Glassware (All Sterile)

Dissecting forceps, straight (Fisher, cat. no. S17327).

Dissecting forceps, blunt (Fisher, cat. no S17326).

Fisherbrand* Mayo dissecting scissors straight (Fisher, cat 13-804-6).

Iris scissors straight (Fisher, cat S17337).

Iris scissors curved (Fisher, cat S17338).

Fisherbrand* delicate dissecting scissors straight (Fisher, cat 08-951-10).

Spatula microdissecting (Fisher, cat NC9005108).

Fisherbrand* Handi-Hold* Microspatula (Fisher, cat 21-401-15).

BD Falcon* Cell strainers, mesh size 70 μm (Fisher, cat 08-771-2).

15-mm petri dish (bacterial grade, non TC-treated; BD Falcon, cat. no. 351029).

Conical tubes 15 and 50 ml.

Glass slides Cat #15-188-52 size 3″ × 1″ Fisher Bioteck probon +.

70 % ethanol.

4 % paraformaldehyde (PFA) in 1× PBS.

37 °C, 5.0 % CO_2 incubator, 95 % humidity.

Biosafety laminar flow/tissue culture hood.

2.4 Preparation of Anti-PSA-NCAM-Coated Dishes

The surface of 100 mm non-tissue culture grade petri dishes coat with 4 ml of the immunopanning cocktail (50 mM Tris–Cl, pH 9.5/1 % BSA (w/v)/50 μg/ml anti-PSA-NCAM antibody). Remove cocktail after 40 min and wash petri dishes three times with PBS, then once with 5 ml PBS/1 % BSA before use.

3 Methods

3.1 Initiation (Fig. 1)

1. Place the brain tissue in a 50-ml conical tube. Carefully remove meninges. Wash with 20 ml cold Hibernate E medium with pen/strep by gently inverting the tube. Repeat three times—7–10 min.
2. Incubate brain tissue in the presence of Tryple™ Express… and DNase I (10 U/ml; Sigma, St. Louis, MO) 10 min at 37 °C—10 min.
3. Add 4 ml of neurobasal medium to the brain tissue and mechanically dissociate the tissue by gently aspirating and releasing the suspension. Avoid the formation of bubbles and foaming—5 min.
4. Triturate cells through a fire-polished Pasteur pipette and filter the cell suspension through a cell strainer (mesh size 70 μm) to remove cell clusters—5 min.
5. Add 10 ml of neurobasal medium. Centrifuge cells for 7 min at 500 × *g*, room temperature—8–10 min.
6. Discard the supernatant. Resuspend the cell pellet in 10 ml of neurobasal medium and gently dissociate the pellet. Transfer the half of the cell suspension into new tube (tube N—for preparation of neurons). Another half (tube NPC) will be used for preparation of glial progenitor cells—3 min.

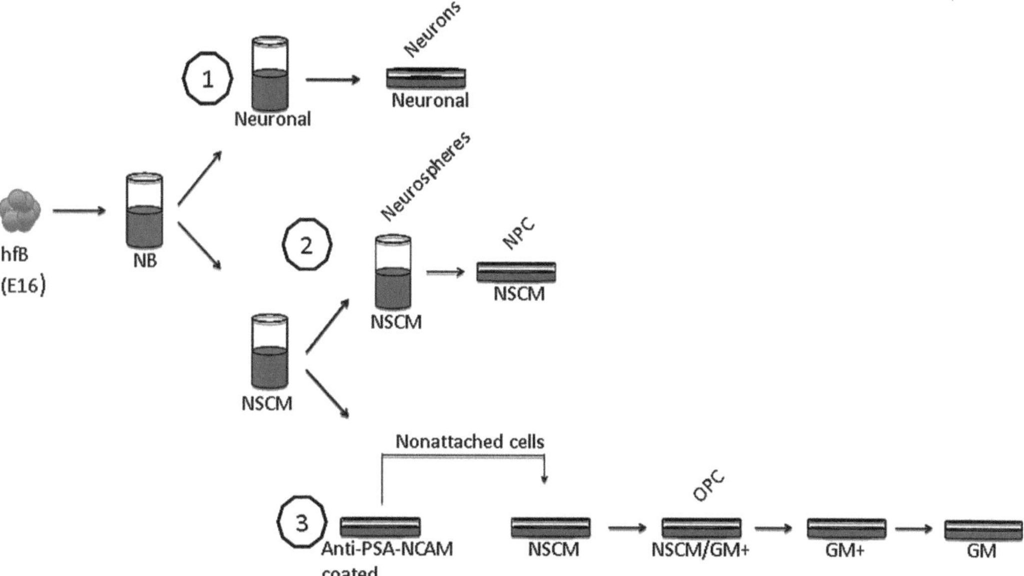

Fig. 1 Schematic presentation of initiation and preparation of neural cultures: 1. Preparation of neurons. 2. Propagation of neuronal stem cells. 3. Differentiation of neural progenitor cells into oligodendrocyte lineage

3.2 Isolation and Expansion of Primary Human Embryonic Neural Progenitor Cells (hNPC) in Chemically Defined Media (Fig. 2)

1. Assess cell viability using Trypan Blue stain (cells with damaged plasma membrane are stained blue).
2. Count cells using hemacytometer and plate onto fresh PSA/NCAM–coated dishes (1×10^6 cells/100-mm) dish in 7 ml of NSCM.
3. Incubate cells overnight at 37 °C with 5.0 % CO_2 and 95.5 % humidity.
4. Next day, transfer nonattached cells to a 15 ml conical tube, centrifuge for 7 min at $500 \times g$, room temperature, save the conditioned medium .
5. Dissociate pellet in NSCM and plate cells onto fresh PSA/NCAM–coated dishes
6. Incubate cells overnight at 37 °C with 5.0 % CO_2 and 95.5 % humidity.
7. On the next day, transfer nonattached cells to a 15 ml conical tube, centrifuge for 7 min at $500 \times g$. Dissociate pelleted cells in 5 ml of their own medium and in 5 ml of NSCM and plate on poly-D-Lysine coated dishes/slides at a density of approximately 1.5×10^6 per 60 mm dish for experiments or further differentiation into oligodendrocyte lineage. Feed cultures with fresh NSCM and one-third conditioned NSCM every other day supplemented with 10 ng/ml human bFGF, 10 ng/ml recombinant human EGF, and 10 ng/ml recombinant human IGF-1 (*see* **Note 2**).

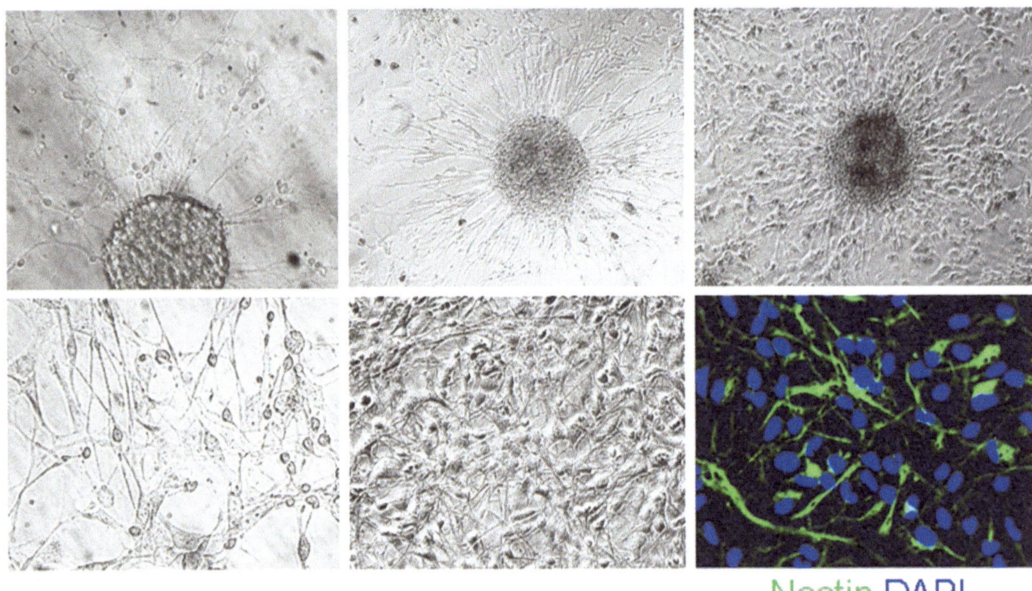

Fig. 2 Isolation and expansion of primary human embryonic neural progenitor cells (hNPC) in chemically defined media

8. Cells also can be propagated in flasks as three dimensional cultures of "neurospheres." Neurospheres, which grow at a slower pace than attached cells, can serve as a renewable source of neural stem cells. Neurosphere cultures grown in flasks should be dissociated routinely to keep the spheres uniformly small (approximately 2 mm in diameter).

3.3 Differentiation of hNSC into the Oligodendrocyte Lineage (Fig. 3)

The hOPC arise from the neural precursors with a characteristic bipolar shape and undergo sequential morphological rearrangements during the course of differentiation as they acquire features specific to OL [6]. Each developmental stage of OL is characterized by the presence of specific markers. Neural stem cells express NKX2.2. PSA-NCAM, Nestin, Pax6. OPCs, while continuing to express Nkx2.2, become positive for Olig1, Olig2, NG2, A2B5, CD3, Sox9, and PDGFRα. Immature Ol express O4, O1, 2′,3′-cyclic nucleotide 3′ phosphohydrolase (CNP), followed by galactocerebroside (GalC). Mature Ol in culture are characterized by highly developed network of cytoplasmic extensions and expression of myelin-specific proteins proteolipidprotein (PLP), myelin basic protein (MBP), myelin-associated glycoprotein (MAG), and myelin oligodendrocyte glycoprotein (MOG). Like NPC, oligodendrocyte progenitors (hOPC) can be propagated in two- and three-dimensional cultures (oligospheres) [7]. Like NPC, attached OPC grow faster. The transition of hNPC to commit to the OL lineage is achieved by gradual switch of the culture medium to Glia

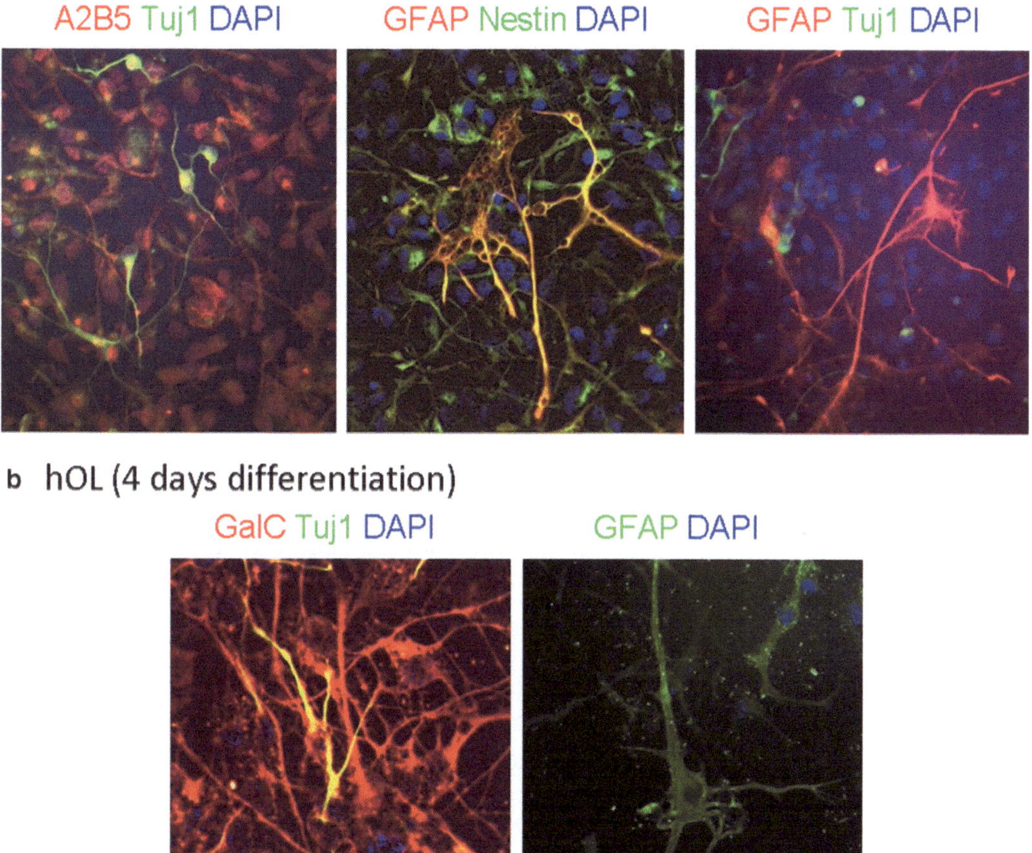

Fig. 3 Differentiation of hNSC into the oligodendrocyte lineage. (**a**) hOPC were immunolabeled for A2B5, Tuj1, GFAP, and Nestin and labeled with DAPI. (**b**) hOL were immunolabeled for GalC, Tuj1, and GFAP and labeled with DAPI

defined medium with growth factors (GM+) To enhance maturation of OL, cells will be cultured in GM without bFGF, EGF, and PDGF-AA supplementation [8]

1. Feed NPC plated onto poly-D-lysine coated dishes with a medium containing one part of self-conditioned NSCM and one part of an oligodendrocyte specification medium (OM). OM containing one part of NSCM and one part of glia defined medium with growth factors (GM+) (DMEM/F12 high glucose, N2 supplement, GlutaMAX, 20 ng/ml bFGF, 20 ng/ml EGF, 20 ng/ml PDGF-AA, and penicillin/streptomycin).
2. The transition of hNPC to commit to the OL lineage is achieved by gradual switch of the culture medium to GM+.
3. To enhance maturation of OL culture cells in GM without bFGF, EGF and PDGF-AA supplementation during 9 days.

4. Characterizing NPC and OPC.
   (a) Immunostaining.
      - Fix the cells with freshly prepared 4 % PFA at room temperature for 7 min (*see* **Note 3**).
      - Gently rinse the cells without dislodging them with 1× PBS.
      - Perform immunostaining according protocols [6, 9].
   (b) Analysis of expression of molecular markers by quantitative polymerase chain reaction (Q-PCR).
      - RNA preparation, Isolate total RNA using the RNeasy kit (Qiagen, Valencia, CA) with on column DNA digestion.
      - Reverse transcription (RT).
      - Perform the RT reaction using 1 μg of total RNA, oligo (dT) 12–18 primer (P/N 58862, Invitrogen) and M-MuLV RT enzyme as we have described previously [10].
      - Quantitative real-time polymerase chain reaction (QPCR). Dilute cDNA samples five times and perform PCR using Roche Applied Sciences Sybr Green mix (cat # 04707516001) on LightCycler480 machine (Roche). PCR reaction conditions: activation 95 °C 5 min, PCR 45 cycles: 95 °C 10 s, 60 °C 20 s, 72 °C 30 s, melting curve (95–65 °C), cool down 40 °C 30 s. The following primers were used:

        PDGFRA,S:5′-TCAGCTACAGATGGCTTGATCC-3′.

        AS: 5′-GCCAAAGTCACAGATCTTCACAAT-3′.

        Olig1 S: 5′-CTAAAATAGGTAACCAGGCGTCTCA-3.

        AS: 5′-CCCGGTACTCCTGCGTGTT-3′.

        Olig2 S: 5′-TGCGCAAGCTTTCCAA.

        GAT-3′; AS: 5′-CAGCGAGTTGGTGAGCATGA-3′.

        Nkx2.2 S: 5′-TCTACGACAGCAGCGACAAC-3′.

        AS: 5′-CTTGGAGCTTGAGTCCTGAG-3′.

        PLP1, S: 5′-AGTCAGGCAGATCTTTGGCG-3′.

        AS: 5′-GACACACCCGCTCCAAAGAA-3′.

        MBP,S:5′-ACTATCTCTTCCTCCCAGCTTAAAAA-3′.

        AS: 5′-TCCGACTATAAATCGGCTCACA-3′.

        Nanog, S: 5′-ATGCCTCACACGGAGACTGT-3′.

        AS: 5′-AAGTGGGTTGTTTGCCTTTG-3′.

        OCT4B, S: 5′-GTTAGGTGGGCAGCTTGGAA-3′.

        AS: 5′-TGTGGCCCCAAGGAATAGTC-3′.

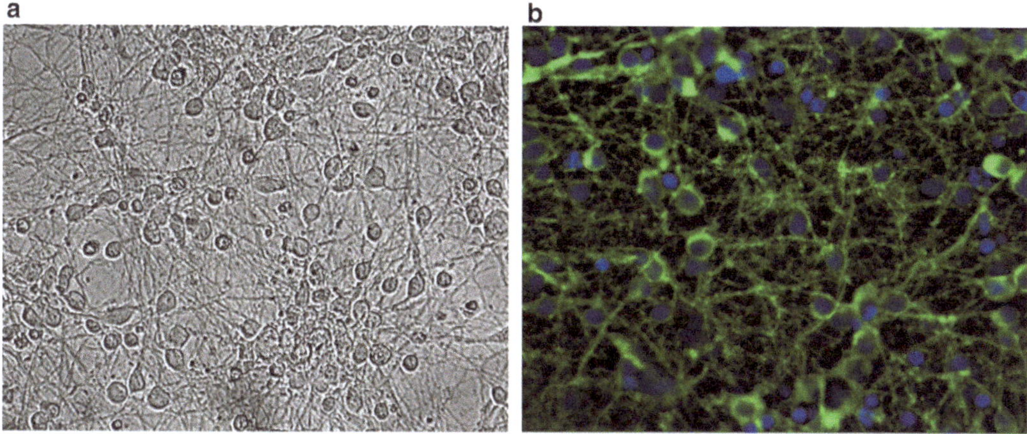

Fig. 4 Human primary neurons (prepared from fetal brain tissue, embryonic age 16 weeks). (**a**) Phase image of live cultures. (**b**) Neurons stained with antibody specific to class III β-Tubulin (*green fluorescence*). Nuclei are stained with DAPI (*blue*)

For relative quantification, the expression level of genes was normalized to the housekeeping gene β-Actin.

β-Actin, S: 5′-CTACAATGAGCTGCGTGTGGC-3′.

AS: 5′-CAGGTCCAGACGCAGGATGGC-3′.

3.4 Isolation and Maintenance of Primary Human Embryonic Cortical Neurons (Fig. 4)

1. Assess cell viability in the tube N using Trypan Blue stain (cells with damaged plasma membrane are stained blue).
2. Count cells using hemacytometer and plate onto fresh poly-D-lysine–coated dishes in neuronal medium (1.8×10^6 cells per 60-mm dish in 4 ml of neuronal medium).
3. After 16 h add Cytosine arabinoside (Ara-C, Sigma; final concentration 1 μM) was added after 16 h for 2 days to reduce glial proliferation.
4. Characterizing neurons:
   - Fix the cells with freshly prepared 4 % PFA at room temperature for 7 min.
   - Gently rinse the cells without dislodging them with 1× PBS
   - Perform immunostaining according previously protocols [6, 9].

4 Notes

1. We do not use antifungal reagents during preparation of cultures from rodents.
2. The final concentration of poly-D-lysine solution for coating plastic dishes is 100 ng/ml, and 500 ng/ml for coating glass slides.

3. We add gradually increasing amounts of fixative to the cells without removing culture medium. When the volume of fixative becomes equal to the volume of culture medium, medium with fixative will be replaced by fixative only and incubated for 5 min at room temperature.

References

1. Espinosa-Jeffrey A, Wakeman DR, Kim SU, Snyder EY, de Vellis J (2009) Culture system for rodent and human oligodendrocyte specification, lineage progression, and maturation. Curr Protoc Stem Cell Biol Chapter 2: Unit 2D.4
2. Pollard SM, Conti L, Sun Y, Goffredo D, Smith A (2006) Adherent neural stem (NS) cells from fetal and adult forebrain. Cereb Cortex 16(Suppl 1):i112–i120
3. McCarthy KD, de Vellis J (1980) Preparation of separate astroglial and oligodendroglial cell cultures from rat cerebral tissue. J Cell Biol 85: 890–902
4. Vicario-Abejón C, Johe KK, Hazel TG, Collazo D, McKay RD (1995) Functions of basic fibroblast growth factor and neurotrophins in the differentiation of hippocampal neurons. Neuron 15:105–114
5. Rowitch DH, Kriegstein AR (2010) Developmental genetics of vertebrate glial-cell specification. Nature 468:214–222
6. Darbinyan A, Kaminski R, White MK, Darbinian-Sarkissian N, Khalili K (2013) Polyomavirus JC infection inhibits differentiation of oligodendrocyte progenitor cells. J Neurosci Res 91(1):116–127. doi:10.1002/jnr.23135
7. Avellana-Adalid V (1996) Expansion of rat oligodendrocyte progenitors into proliferative "oligospheres" that retain differentiation potential. J Neurosci Res 45:558–570
8. Neman J, de Vellis J (2012) A method for deriving homogenous population of oligodendrocytes from mouse embryonic stem cells. Dev Neurobiol 72(6):777–788. doi:10.1002/dneu.22008
9. Merabova N, Kaminski R, Krynska B, Amini S, Khalili K, Darbinyan A (2012) JCV agnoprotein-induced reduction in CXCL5/LIX secretion by oligodendrocytes is associated with activation of apoptotic signaling in neurons. J Cell Physiol 227:3119–3127
10. Merabova N, Kaniowska D, Kaminski R, Deshmane SL, White MK, Amini S, Darbinyan A, Khalili K (2008) JC virus agnoprotein inhibits in vitro differentiation of oligodendrocytes and promotes apoptosis. J Virol 82: 1558–1569

Chapter 6

Mouse Enteric Neuronal Cell Culture

Yonggang Zhang and Wenhui Hu

Abstract

In the enteric nervous system, there exist a lot of local intrinsic neurons which control the gastrointestinal functions. Culture of enteric neurons provides a good model system for physiological, electrophysiological, and pharmacological studies. Here, we describe two methods to obtain sufficient enteric neurons from mouse myenteric plexuses by directly culturing primary neurons or inducing neuronal differentiation of enteric neural stem/progenitor cells.

Key words Enteric neurons, Neural stem/progenitor cells, Neuronal differentiation, Cell culture, Mouse

1 Introduction

The enteric nervous system, also called "the second brain," is embedded within the wall of gut, which control the complex functions of gastrointestinal tract such as peristalsis, blood flow, and secretion [1–3]. The enteric nervous system consists of at least two major ganglionated plexuses, the myenteric plexus between the circular and longitudinal muscle layers and the submucosal plexus between the circular muscle and mucosal layers [2, 3]. The cell bodies of enteric neurons are surrounded and protected by enteric glial cells [4]. Enteric neurons from different plexuses have different functions and many subtypes of enteric neurons exist in the plexuses [5–7].

There are some difficulties in isolating and culturing enteric neurons in contrast to brain neurons: food debris, bacteria, and fungi in the lumen of the gut; mixed cell types in the tissue or plexus preparation; long network between various plexuses; and anatomically diffuse distribution of plexuses embedded into smooth muscle layers and mucosa layers. Most early studies used guinea pig for enteric neuronal culture due to relatively easy isolation of the loose and large plexuses [8–10]. Also, culture of enteric neurons in rat and human has been established [9, 11–13]. Recently, successful culture of mouse enteric neurons has been reported [14, 15]. The mouse model attracts wide attention

Shohreh Amini and Martyn K. White (eds.), *Neuronal Cell Culture: Methods and Protocols*, Methods in Molecular Biology, vol. 1078,
DOI 10.1007/978-1-62703-640-5_6, © Springer Science+Business Media New York 2013

because of the versatility in inbred and genetically engineered strains and the economic cost. In this chapter, we describe a unique isolation of mouse myenteric plexuses, the primary culture of mouse enteric neurons as well as the culture and neuronal differentiation of enteric neural stem/progenitor cells.

2 Materials

2.1 Isolation of Smooth Muscle/Myenteric Plexus (SM/MP) Strips

1. Fine scissors and fine forceps.
2. Stereomicroscope.
3. Agar plate (2 % agar PBS in 10 cm sterile petri dish, *see* **Note 1**).
4. Ice box.
5. Sterile petri dish.

2.2 Reagents

1. Ice cold HBSS with 10 mM HEPES (calcium and magnesium free).
2. Enteric neuronal culture medium (equal to enteric neural stem/progenitor cell differentiation medium):

   To make 50 ml of complete medium:

DMEM/F12 medium	44 ml
Heat inactivated Fetal Bovine Serum (HS-FBS)	1 ml
Chick Embryo Extract	2.5 ml
Penn/Strep (100×)	0.5 ml
Gentamicin (500×)	100 μl
Amphotericin (100×)	0.5 ml
N2 (100×)	0.5 ml
B27 (50×)	1 ml

3. Enteric neural stem/progenitor cell proliferation medium:

   To make 50 ml of medium:

DMEM/F12 medium	39.5 ml
Chick Embryo Extract (CEE)	7.5 ml
Penn/Strep (100×)	0.5 ml
Gentamicin (500×)	100 μl
Amphotericin (100×)	0.5 ml
N2 (100×)	0.5 ml
B27 (50×)	1 ml
Glutamine (100×)	0.5 ml

(continued)

FGF-b (10 μg/ml)	50 μl
EGF (10 μg/ml)	100 μl
Heparin (0.2 %)	5 μl

4. Digestion medium:
   (a) Collagenase digestion medium: 1 mg/ml collagenase IV (from Worthington), 0.5 mM $CaCl_2$, 10 mM HEPES in HBSS.
   (b) Trypsin digestion medium: 0.05 % Trypsin with 0.53 mM EDTA in HBSS.
   (c) Digestion neutralizing medium: 400 U DNase, 1 mg/ml BSA in DMEM/F12 medium.
5. Growth factor-reduced Matrigel.

3 Methods

3.1 Isolation of Smooth Muscle/Myenteric Plexus (SM/MP) Strips from Mouse Gut

(The whole procedure is done on ice, *see* **Note 2**)

1. Immediately prior to the experiment, euthanize the adult mouse using CO_2.
2. Dissect out the whole small intestine (usually from duodenum to ileum or the regions of interest), and put into 10 cm petri dish filled with ice-cold HBSS (Ca^{2+} and Mg^{2+} free) solution.
3. Carefully remove mesenteries, fat tissue, and gut-associated organs free from the gut wall.
4. Flush the stool from ileum to duodenum three times using 10 ml syringe connected with a polyethylene tube.
5. Wash the gut with HBSS three times and transfer into a new sterile petri dish containing HBSS solution.
6. Cut the gut into 8 cm long pieces and use the scissors to open the lumen.
7. Transfer one piece into agar plate (*see* **Note 1**). Use forceps to unfold the gut with the lumen upward for SM/MP dissection. At one end of the gut, use one forceps to hold the gut and the other one to scratch off the mucosa layer followed by submucosa layer carefully under the stereomicroscope to make a window. When the submucosa layers are separated from SM/MP layers, use one forceps to hold the SM/MP layers and the other forceps to peel off the submucosa and mucosa layer (*see* **Note 3**).
8. Transfer the dissected SM/MP strips into new sterile petri dish containing HBSS.
9. Use the fine scissors to cut the layers/MP strips into about 1 cm long pieces and transfer all of them into 15 ml tube.

3.2 Enzymatic Dissociation of SM/MP Tissues

1. Centrifuge the collected tissues (200 × *g*, 5 min at 4 °C) and aspirate the solution, add 3 ml prewarmed (37 °C) collagenase digestion medium to resuspend the tissue and incubate in 37 °C water bath shaker (40 rpm) for 15 min (*see* **Note 4**). For every 5 min, take the 15 ml tube out and invert several times to resuspend the tissues.
2. Release the smooth muscle cells from the myenteric plexuses (*see* **Note 5**):

   Spin down the dissociated tissue (200 × *g*, 5 min at 4 °C) and aspirate the solution. Add 2 ml HBSS and use a cut-off P1000 pipette tip to transfer the pellet into a 10 cm petri dish containing 10 ml HBSS. Shake softly (15 rpm) for 10 min at room temperature. Use the sterile forceps to transfer the tissues into a new 10 ml petri dish containing 10 ml HBSS and shake 10 min at room temperature.
3. Transfer all the tissues into 15 ml tube with 1 ml of Trypsin digestion medium, and incubate in 37 °C water bath shaker (40 rpm) for 10 min.
4. Add 2 ml digestion neutralizing medium into the tube and spin down at 200 × *g*, 5 min, 4 °C. Aspirate the solution, add 1 ml digestion neutralizing medium and triturate vigorously using a P1000 pipette until no visible tissue clumps remain.
5. Add the HBSS solution to 10 ml, spin down the cells (200 × *g*, 5 min at 4 °C), and resuspend the cells using 1 ml digestion neutralizing medium.
6. Count the cell number using standard Trypan blue staining and hemocytometry.

3.3 Culture of Primary Enteric Neurons

1. Matrigel coating of the culture plates (*see* **Note 6**):

   Thaw the Matrigel on ice (If at room temperature, it will form the gel quickly).
   - Dilute the Matrigel (1:200) in ice cold DMEM/F12 medium, vortex 10 s. Precool the pipette tip before transferring the Matrigel
   - Add the diluted Matrigel into the precooled culture plates and put into the incubator (37 °C) for 1–2 h. The coating volume used depends upon the experimental purpose: for 10 cm dish, 6 ml; 60 mm dish, 3 ml; 35 mm dish or 1 well of 6-well plate, 1.5 ml; 24-well plate, 0.5 ml/well; 96-well-plate or 8-well chamber slide, 100 μl/well. For immunocytochemistry and/or confocal imaging, sterile coverslips can be added into 6-well or 24-well plate.
   - Wash the coated wells once with HBSS, and add half volume of the culture medium immediately (the other half will be added together with the cells). Do not let the dish or plate dry after coating.

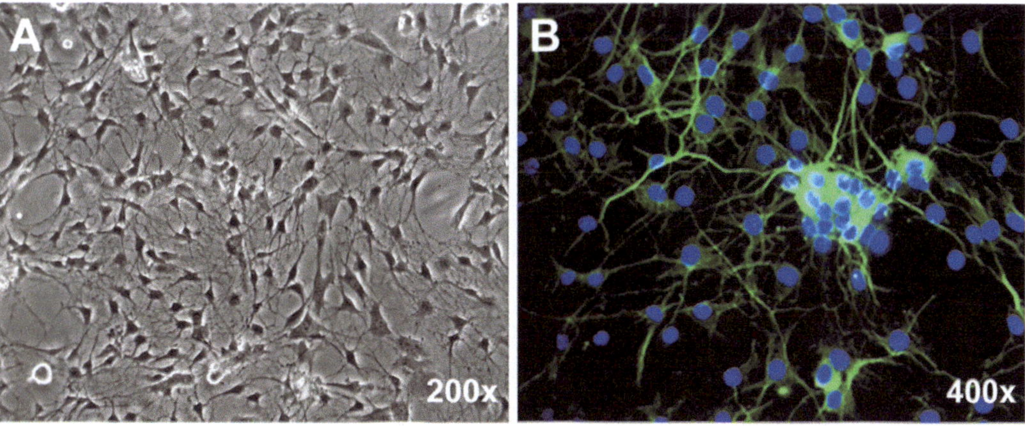

Fig. 1 Representative phase-contrast (**a**) and immunofluorescent (**b**) micrographs of primary enteric neurons cultured from small intestinal myenteric plexuses of 8-week-old mouse. After 5 days in culture, cells were fixed with 4 % paraformaldehyde/PBS for 10 min and standard immunocytochemistry was performed using rabbit anti-Tuj-1 polyclonal antibody (1:5,000, Sigma) and Alexa Fluor® 488 (*green*) conjugated donkey anti-rabbit secondary antibody (1:400; Invitrogen). Hoechst 33258 was used for counterstaining of nuclei (*Blue*)

2. Seed the dissociated cells into Matrigel-coated dish or plate at the density of $1 \times 10^5/cm^2$ with enteric neuronal culture medium, and culture at 37 °C with 5 % CO_2, change medium every day.
3. The morphological changes of the cultured cells are observed daily under phase-contrast microscope. After 1–2 days in culture, enteric neurons attach to the coated surface and develop long processes. After 5–7 days in culture, some neurons attempt to form ganglion-like plexuses over a layer of the flatter, wider glial cells (Fig. 1a). Immunocytochemical staining using neuronal marker Tuj-1 (β tubulin III) validates the success of enteric neuronal culture (Fig. 1b).

3.4 Culture and Differentiation of Enteric Neural Stem/Progenitor Cells (See Note 7)

1. Primary neurosphere culture of enteric stem/progenitor cells.

   Seed the dissociated cells with the enteric neural stem/progenitor cell proliferation medium at the density of $1 \times 10^5/cm^2$ in 60 mm sterile petri dish without any coating (5 ml culture medium per dish), and culture at 37 °C with 5 % CO_2. After 6–7 days in culture, the enteric neurosphere will be ready for passage (Fig. 2a). Around 1,500–2,000 enteric neurospheres per adult mouse can be obtained.
2. Dissociation of enteric neurospheres into single cells.
   - Warm enteric neural stem/progenitor cell proliferation medium to 37 °C. Collect primary neurospheres to 15 ml tube, and spin down the neurospheres by centrifuge at 400 rpm ($90 \times g$) for 1 min.
   - Aspirate the supernatant and add the Accutase solution to resuspend the neurospheres. Incubate at 37 °C for 15 min. Here is one example for neurospheres cultured in

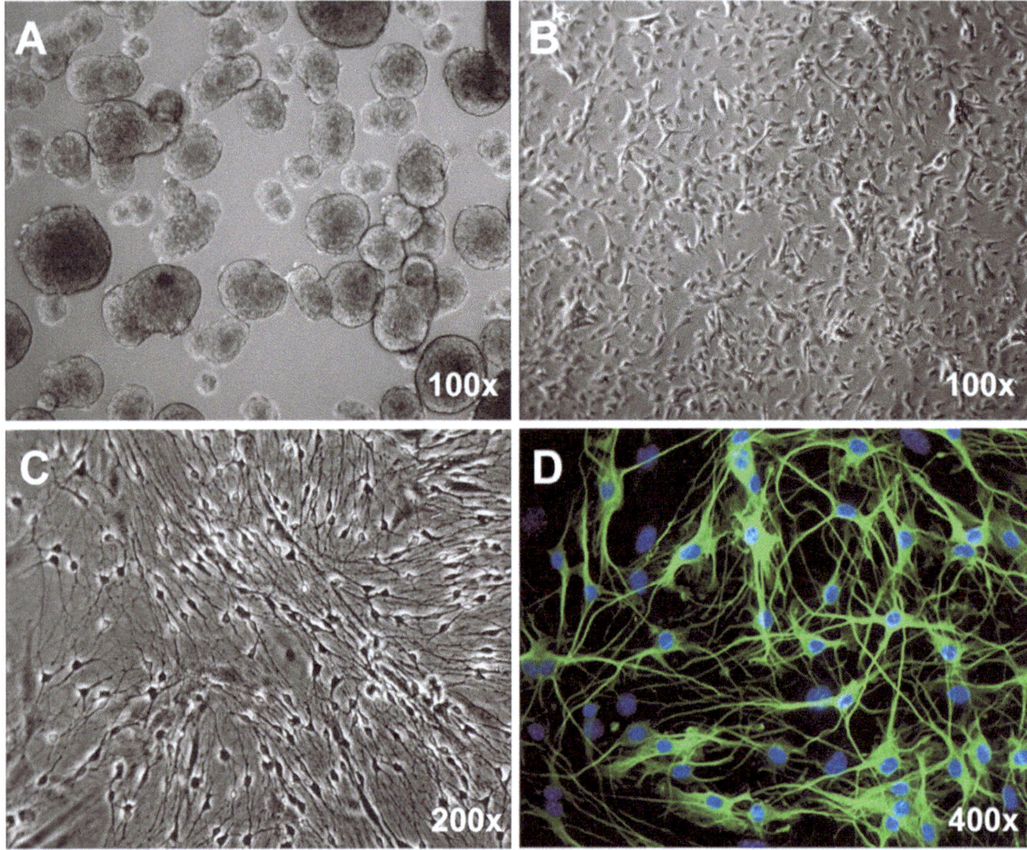

Fig. 2 Culture and differentiation of enteric neural stem/progenitor cells from 8-week-old mouse. (**a**) Neurosphere culture for 7 days. (**b**) Monolayer culture. (**c**, **d**) Differentiation for 4 days. Representative phase-contrast (**a**–**c**) and immunofluorescent (**d**) micrographs are shown. Standard immunocytochemistry was performed using rabbit anti-Tuji1 polyclonal antibody (1:5,000, Sigma) and Alexa Fluor® 488 (*green*) conjugated donkey anti-rabbit secondary antibody (1:400; Invitrogen). Hoechst 33258 was used for counterstaining of nuclei (*Blue*)

a 60-mm dish using 0.8 ml Accutase solution. Alternatively, the collagenase IV at final concentration of 1 mg/ml also works fine.

- Add 2× volumes of DMEM/F12 medium to 15 ml tube to dilute the enzyme. Centrifuge at 210 × *g* for 4 min.
- Aspirate off all of the supernatant and resuspend cells in 0.2 ml neural stem/progenitor cell proliferation medium by gently pipetting the cell pellet using the 200 μl tip, and then add the medium to 1 ml. Count the cell number using standard Trypan blue staining and hemocytometry.

3. Expansion of enteric neural stem/progenitor cells (*see* **Note 8**).

   Seed the dissociated single cells on the Matrigel-coated plate at the density of 2×10^4 cells/cm^2 and culture with the enteric neural stem/progenitor cell proliferation medium at 37 °C with 5 % CO_2. Change medium every day. At 90–95 %

confluence (Fig. 2b), the cells can be dissociated into single cells and subculture again (1:3).

4. Neuronal differentiation of enteric neural stem/progenitor cells.

   Seed the dissociated single cells on the Matrigel-coated plate or coverslips at the density of 2×10^4 cells/cm^2 and culture with enteric neuronal culture medium at 37 °C with 5 % CO_2. Change medium every day. After 3–4 days in culture, the enteric neurons enriched with neurites will be observed under phase-contrast microscope (Fig. 2c). Immunocytochemical staining using neuronal marker Tuj-1 (β-tubulin III) validate the successful culture of enteric neurons differentiated from enteric neural stem/progenitor cells (Fig. 2d).

4 Notes

1. Smooth muscle/myenteric plexus layer is a thin out layer of gut. It is very soft and easily damaged during the dissection. Agar plate is soft and will avoid damage to tissue during the dissection
2. There are a lot of enzymes in the gut, and the enteric neurons will be easily damaged. So every step of the dissection must be performed on ice. When dissected under the stereomicroscope, solution should be changed every 5 min to keep lower temperature.
3. The myenteric plexus lies between the circular and longitudinal muscle, so we first peel off the mucosa and sub-mucosa layers to reduce the contamination from the contents in the lumen.
4. The time of digestion with collagenase is very critical, and different for individual animals. We checked the digestion every 5 min. When the color of tissue begins to change into white, stop immediately.
5. Smooth muscle/myenteric plexus layer contains a lot of smooth muscle cells. We first used the collagenase to separate the myenteric plexus from two smooth muscle layers and then treated the myenteric plexus with trypsin. After digestion of the smooth muscle/myenteric plexus, the cell suspension is composed of several cell types including neurons, glia cells, smooth muscle cells, fibroblast, and neural stem/progenitor cells. During the culture, smooth muscle cells will die off and the neuron and glial cells will survive due to the culture conditions.
6. We tested different coating matrixes. A few neurons will attach on the poly-D-lysine coated plate. Better adherence can be reached with the poly-D-lysine/fibronectin or poly-D-lysine/laminin coating. However, double coating costs a lot and

needs 1–2 days for the coating process. BD Matrigel™ matrix is a reconstituted basement membrane preparation from the Engelbreth-Holm-Swarm (EHS) mouse sarcoma. This material is comprised of approximately 60 % laminin, 30 % collagen IV, and 8 % entactin. We found that Matrigel-coated dishes or coverslips efficiently support the attachment and growth of primary neurons, differentiating neurons and neural stem/progenitor cells. The coating procedure takes less than 1 h.

7. The enteric neuron is a type of terminally differentiated cell. Direct culture of primary enteric neurons is often difficult to obtain large number of enteric neurons. In vitro culture of enteric neural stem/progenitor cells from embryonic, neonatal, and adult gut tissues has been established [16–20]. Differentiation of cultured enteric neural stem/progenitor cells provides an easy approach to obtain a larger number of enteric neurons which are usually more pure.
8. The enteric neural stem/progenitor cells can be expanded through either monolayer or neurosphere culture. We prefer the monolayer culture because of its higher efficiency of expansion, easy dissociation into single cells, and even exposure to nutrients.

References

1. Wood JD (2010) Enteric nervous system: sensory physiology, diarrhea and constipation. Curr Opin Gastroenterol 26(2):102–108
2. Benarroch EE (2007) Enteric nervous system: functional organization and neurologic implications. Neurology 69(20):1953–1957
3. Brehmer A (2006) Structure of enteric neurons. Adv Anat Embryol Cell Biol 186:1–91
4. Grundy D, Schemann M (2007) Enteric nervous system. Curr Opin Gastroenterol 23(2): 121–126
5. Sternini C (1988) Structural and chemical organization of the myenteric plexus. Annu Rev Physiol 50:81–93
6. Gulbransen BD, Sharkey KA (2009) Purinergic neuron-to-glia signaling in the enteric nervous system. Gastroenterology 136(4):1349–1358
7. Mazzuoli G, Schemann M (2012) Mechanosensitive enteric neurons in the myenteric plexus of the mouse intestine. PLoS One 7(7):e39887
8. Saffrey MJ, Bailey DJ, Burnstock G (1991) Growth of enteric neurones from isolated myenteric ganglia in dissociated cell culture. Cell Tissue Res 265(3):527–534
9. Jessen KR, Saffrey MJ, Burnstock G (1983) The enteric nervous system in tissue culture. I. Cell types and their interactions in explants of the myenteric and submucous plexuses from guinea pig, rabbit and rat. Brain Res 262(1):17–35
10. Hanani M, Xia Y, Wood JD (1994) Myenteric ganglia from the adult guinea-pig small-intestine in tissue-culture. Neurogastroenterol Motil 6(2):103–118
11. Pouokam E, Rehn M, Diener M (2009) Effects of H2O2 at rat myenteric neurones in culture. Eur J Pharmacol 615(1–3):40–49
12. Kugler EM et al (2012) Activity of protease-activated receptors in primary cultured human myenteric neurons. Front Neurosci 6:133
13. Willard AL, Nishi R (1985) Neurons dissociated from rat myenteric plexus retain differentiated properties when grown in cell culture. II. Electrophysiological properties and responses to neurotransmitter candidates. Neuroscience 16(1):201–211
14. Kang SH, Vanden P (2003) Berghe, and T.K. Smith, *Ca2+-activated Cl- current in cultured myenteric neurons from murine proximal colon.* Am J Physiol Cell Physiol 284(4): C839–C847
15. Hagl CI et al (2012) Enteric neurons from postnatal Fgf2 knockout mice differ in neurite outgrowth responses. Auton Neurosci 170(1–2):56–61
16. Lindley RM et al (2009) Properties of secondary and tertiary human enteric nervous system

neurospheres. J Pediatr Surg 44(6):1249–1255, discussion 1255–1256
17. Metzger M et al (2009) Expansion and differentiation of neural progenitors derived from the human adult enteric nervous system. Gastroenterology 137(6):2063–2073, e4
18. Almond S et al (2007) Characterisation and transplantation of enteric nervous system progenitor cells. Gut 56(4):489–496
19. Silva AT et al (2008) Neural progenitors from isolated postnatal rat myenteric ganglia: expansion as neurospheres and differentiation in vitro. Brain Res 1218:47–53
20. Schafer KH, Micci MA, Pasricha PJ (2009) Neural stem cell transplantation in the enteric nervous system: roadmaps and roadblocks. Neurogastroenterol Motil 21(2): 103–112

Chapter 7

Preparation of Neural Stem Cells and Progenitors: Neuronal Production and Grafting Applications

Joseph F. Bonner, Christopher J. Haas, and Itzhak Fischer

Abstract

Neural stem cells (NSC) are not only a valuable tool for the study of neural development and function, but an integral component in the development of transplantation strategies for neural disease. NSC can be used to study how neurons acquire distinct phenotypes and how the reciprocal interactions between neurons and glia in the developing nervous system shape the structure and function of the central nervous system (CNS). In addition, neurons prepared from NSC can be used to elucidate the molecular basis of neurological disorders as well as potential treatments. Although NSC can be derived from different species and many sources, including embryonic stem cells, induced pluripotent stem cells, adult CNS, and direct reprogramming of non-neural cells, isolating primary NSC directly from rat fetal tissue is the most common technique for preparation and study of neurons with a wealth of data available for comparison. Regardless of the source material, similar techniques are used to maintain NSC in culture and to differentiate NSC toward mature neural lineages. This chapter will describe specific methods for isolating multipotent NSC and neural precursor cells (NPC) from embryonic rat CNS tissue (mostly spinal cord). In particular, NPC can be separated into neuronal and glial restricted precursors (NRP and GRP, respectively) and used to reliably produce neurons or glial cells both in vitro and following transplantation into the adult CNS. This chapter will describe in detail the methods required for the isolation, propagation, storage, and differentiation of NSC and NPC isolated from rat spinal cords for subsequent in vitro or in vivo studies.

Key words Neural stem cells, Neural precursor cells, Pluripotency, Multipotency

1 Introduction

1.1 Preface

Neural stem cells (NSC) grown in culture are a valuable tool for the study of neural development, neural function, and the development of therapeutic strategies for neural disease. In the fields of neural development and function, NSC can be used to study how neurons acquire diverse, region specific, phenotypes and how interaction between neurons and glia in the developing nervous system shape the structure and function of the CNS. Although NSC are often considered as a therapeutic tool, they can be used as a source of neurons for mechanistic studies that require a population of naïve neurons that have not yet been influenced by the

Shohreh Amini and Martyn K. White (eds.), *Neuronal Cell Culture: Methods and Protocols*, Methods in Molecular Biology, vol. 1078, DOI 10.1007/978-1-62703-640-5_7, © Springer Science+Business Media New York 2013

CNS environment [1, 2]. NSC can also be used in the study of neurological disorders to elucidate the molecular basis of the disease and how it could potentially be treated. This application has become particularly exciting with the recent progress of obtaining NSC derived from patient cells through the induced pluripotent technology [3, 4].

Although NSC can be derived from different species and many sources, including embryonic stem cells (ESC), induced pluripotent stem cells (iPSC), adult CNS, and direct reprogramming of non-neural cells [5], isolating primary NSC directly from fetal tissue is an important technique with certain advantages. Whether NSC are derived from fetal tissue or pluripotent sources, such as ESC and iPSC, similar techniques are used to maintain NSC in culture and to differentiate NSC toward mature neural lineages. ESC and iPSC cultures produce neurons only after progressing through a sequence of intermediate cell types with concomitant gradual fate restriction, which include the generation of NSC [6]. In this way, pluripotent stem cell research can follow protocols developed for primary NSC culture. This is particularly evident in the derivation of neurons from NSC, where the diversity and complexity of neuronal phenotypes can be difficult to replicate in vitro. This chapter will describe specific protocols for isolating multipotent neuroepithelial cells (NEP) and neural precursor cells (NPC) from embryonic rat CNS tissue (mostly spinal cord). Several laboratories have previously shown that these cell types can be used to reliably produce neurons in vitro [7–9], with NPC also reliably producing neurons after transplantation into the adult CNS [9, 10].

1.2 Terminology

Stem cell research exists at the junction of development, embryology, and cell biology. While this diversity leads to exciting discoveries across multiple areas of research, confusion often arises regarding terminology. Several terms will be briefly defined and discussed for this Chapter. Stem cells are defined by their phenotypic potency and capacity for self-renewal. *Potency* is the range of cells that a particular stem cell can produce and is best tested by clonal analysis. Pluripotent cells can form all the cells of the mature organism, including cells from all 3 germ layers, and theoretically have unlimited capacity for self-renewal. The term embryonic stem cell (ESC) specifically refers to pluripotent stem cells derived from the inner cell mass of a developing embryo. Multipotent cells can produce multiple types of cells but are generally limited to one germ layer. Both their potency and capacity for self-renewal are more limited than pluripotent cells. The term neural stem cells or NSC generally refers to multipotent cells capable of generating neurons, astrocytes, and oligodendrocytes. The terms *precursor* and *progenitor* refer to a class of stem cells that have committed a more restricted lineage profile than NSC, and therefore have a limited potency and limited capacity for self-renewal. Specific examples of NPC present in the developing CNS include neuronal and glial restricted precursors (NRP and GRP, respectively). The term *fetal stem cell* is sometimes used to describe a stem cells derived from developmental tissue beyond the blastocyst stage (i.e., cells more mature than ESC). This is often a misnomer, especially when applied to rodent systems that have a relatively short fetal stage (embryonic days 17–21 in rats); however some human neural stem/progenitor cells derived from aborted tissue can be accurately described as fetal stem cells. *Embryonic day* is a term that is used to describe the age of developing organisms. For the purposes of harvesting stem cells, the embryonic day is defined as the number of days since mating/fertilization. However, disagreement exists over how fertilization should be assigned a numerical value. Fertilization is alternatively defined as the beginning of embryonic day 0 (E0) or the beginning of the embryonic day 1 (E1). Likewise, fertilization + 12 h is alternatively defined as E0.5 or E1.5. For the purposes of this chapter we will continue to use the nomenclature that we have used in our previous studies, with fertilization being defined as the beginning of E0; however, readers should note that many animal vendors designate fertilization as E1 when selling timed pregnant rats.

1.3 Neural Stem Cells and Neural Progenitors of the Developing Spinal Cord

The developing rat spinal cord has been well described in the literature, both in terms of isolation of NSC and NPC populations for in vitro studies, as well as the use of NSC, NPC, and fetal spinal cord tissue for transplantation studies in vivo [11–13]. At E10.5 the caudal neural tube contains neuroepithelial cells (NEP), a multipotent NSC population capable of generating neurons, astrocytes, and oligodendrocytes. Clonal analysis has confirmed that NEP are a common progenitor for the cellular phenotypes found in the adult spinal cord [14], including motor neurons, interneurons, and sensory neurons [7]. NEP can be identified by immunocytochemistry for Nestin (Fig. 1a) and Sox2. Although NEP have a robust capacity to generate multiple neuronal and glial phenotypes in vitro, they show poor survival following transplantation into the adult CNS, likely due to a lack of trophic support in the mature CNS microenvironment [10]. Other sources of multipotent NSC show improved survival, but will only produce glial progeny in the spinal cord, despite being able to produce neurons in neurogenic regions on the CNS [15, 16]. Thus, NEP are a useful source of multiple neuronal phenotypes for in vitro studies, but must be predifferentiated to a more mature, intermediate stage (e.g., NRP) prior to use in transplantation studies.

Later in development, at E13.5, the spinal cord contains both neuronal and glial restricted precursors (NRP and GRP, respectively). NRP will only differentiate into neurons [17] and GRP will only differentiate into astrocytes and oligodendrocytes [18]. Thus NRP/GRP cultures are a source for both neurons and glia (Fig. 2). NRP can be derived from ESC, NEP, or directly from the developing spinal cord and retain the ability to generate multiple neurotransmitter phenotypes both in vitro and in vivo [19, 20]. Like NEP, NRP/GRP express the intermediate filament Nestin (Fig. 1b), but NRP can be specifically identified by ENCAM

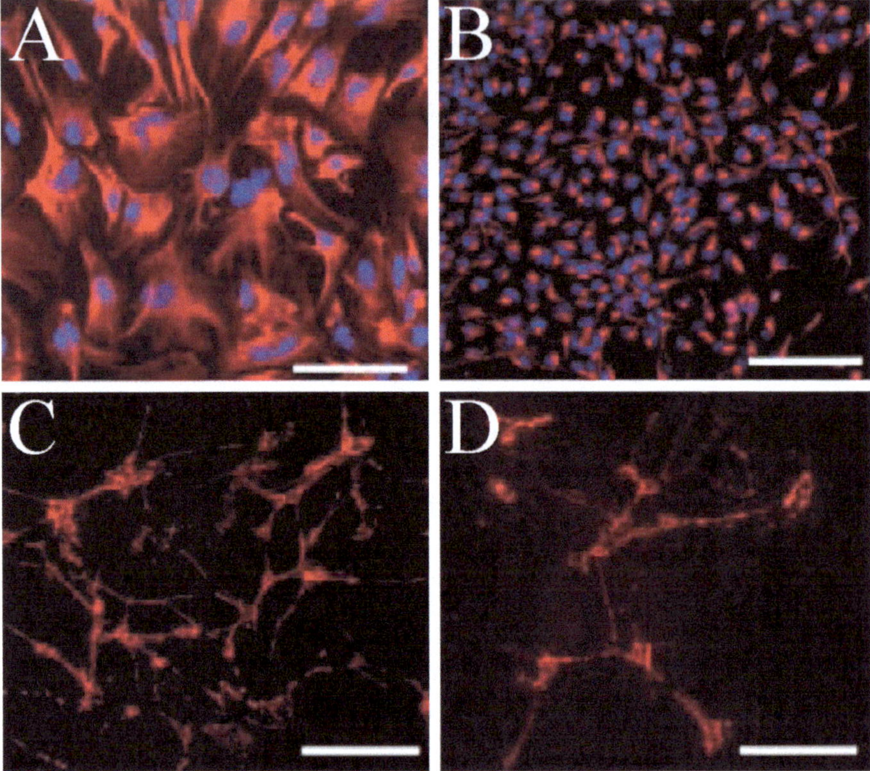

Fig. 1 Neuroepithelial cells (NEP) grow on fibronectin substrates in the presence of bFGF and CEE express the intermediate filament, nestin (**a**, *red*), a neural stem cell marker, throughout the cell body. NRP/GRP grown on poly-L-lysine and laminin substrates in the presence of bFGF and NT-3 also express nestin (*red*, **b**). However, NRP also express ENCAM (**c**) and GRP also express A2B5 (**d**). ENCAM and A2B5 can be used to specifically identify NRP and GRP, respectively, from primary E13.5 spinal cord cultures. Scale bars = 50 μm

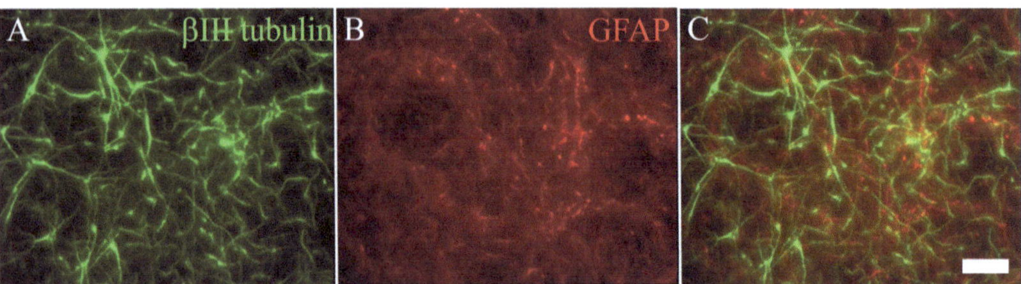

Fig. 2 NRP/GRP grown on PLL/Laminin substrate produce high density βIII tubulin + neurons (**a**) and GFAP+ astrocytes (**b**) when grown in NRP basal medium supplemented with 0.5 % High concentration Matrigel (BD Bioscience, San Jose, CA)

expression (Fig. 1c), whereas GRP can be specifically identified by A2B5 expression (Figs. 1d and 3a). Unlike NEP, NRP survive transplantation in the adult CNS. When transplanted into the intact CNS, NRP will survive and generate neurons [9] (Fig. 4) with appropriate neurotransmitter identities (Fig. 5), per the local

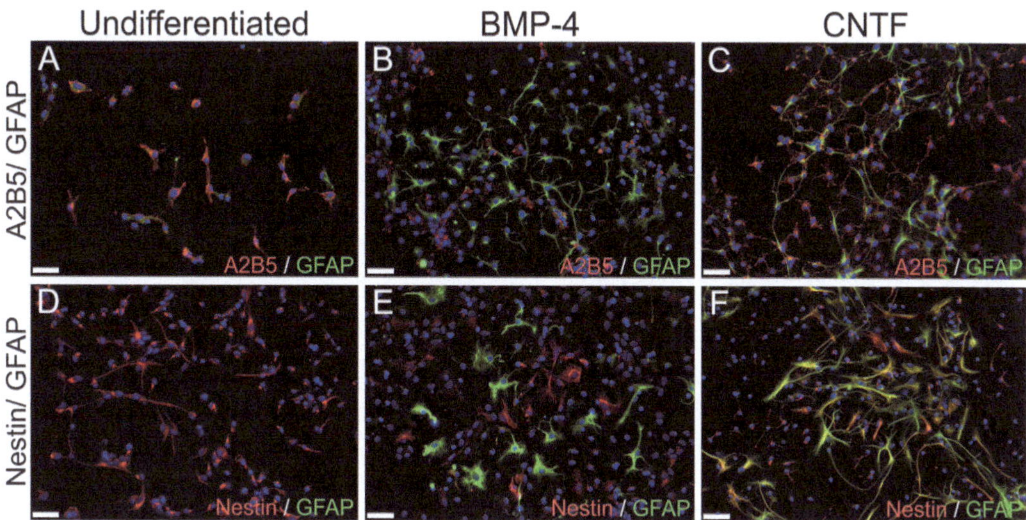

Fig. 3 Enriched GRP grown on PLL substrate express high levels of A2B5 (**a**) and Nestin (**d**). When differentiated with BMP-4 (**b**, **e**) or CNTF (**c**, **f**) GRP produce astrocyte subpopulations characterized by low levels of A2B5 and Nestin, but high levels of GFAP (BMP-4), or intermediate levels of A2B5 and Nestin, and high levels of GFAP (CNTF). Scale bar = 25 μM

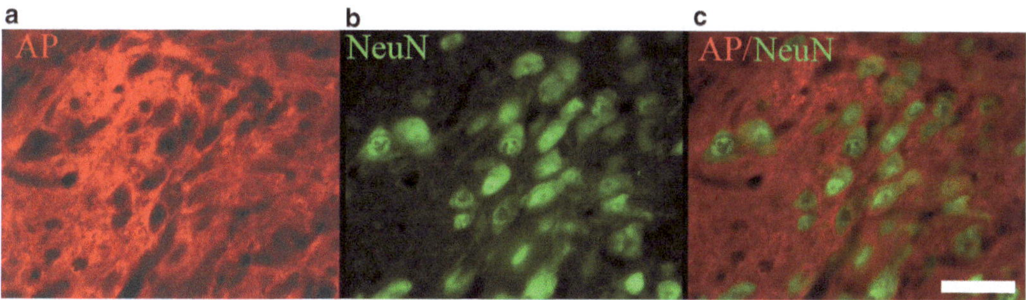

Fig. 4 NRP/GRP derived from AP+ rat embryos survive in the injured cervical spinal cord and produce neurons after differentiation. Immunohistochemistry demonstrates that NRP/GRP express the AP transgene (**a**, **c**) and produce NeuN+ neurons (**b**, **c**) 4 weeks after grafting. Scale bar = 50 μm

microenvironment [19]. However, when purified NRP are transplanted into the injured CNS, their survival and differentiation is poor because the injured spinal cord lacks the appropriate microenvironment to support NRP [8]. We have previously demonstrated that combined grafts of NRP/GRP in the injured spinal cord produce astrocytes, oligodendrocytes, and neurons, indicating that GRP are capable of producing a microenvironment that supports NRP survival and differentiation [21]. NRP will produce both glutamatergic and GABAergic neurons in the injured spinal cord when grafted with GRP. Neurons derived from NRP have been shown to form synapses in vitro [17] and in vivo [22], demonstrating that NRP produce neurons with characteristics of mature, functional neurons.

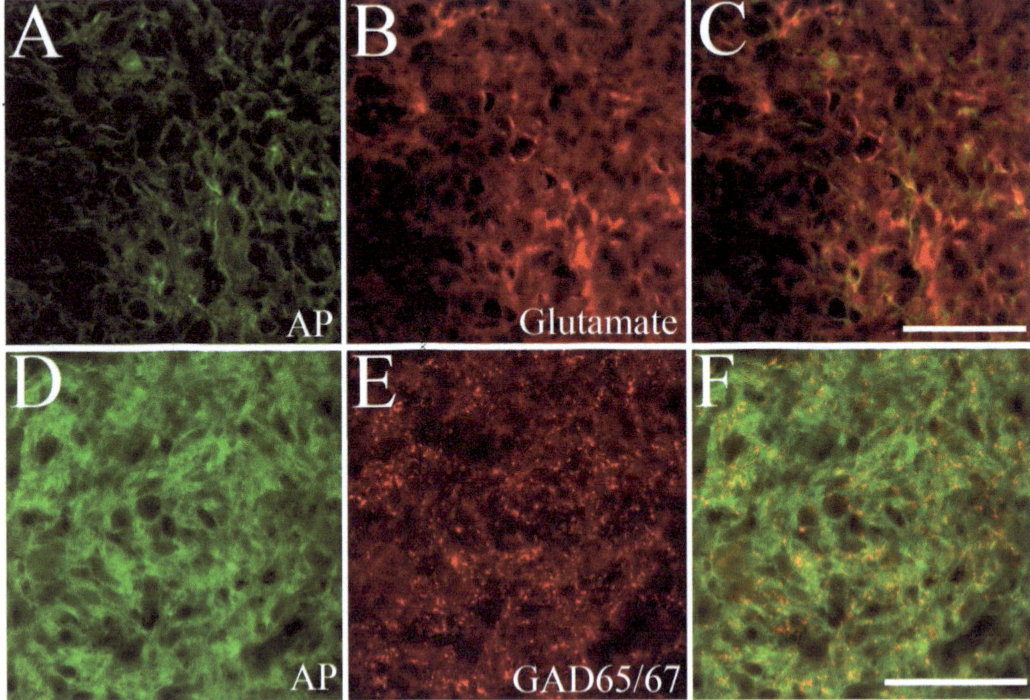

Fig. 5 NRP/GRP derived from AP+ rat embryos were grafted into the injured cervical spinal cord of adult rats. NRP/GRP survive, fill the lesion, and continue to express the AP transgene 6 weeks after grafting in the injured spinal cord (**a**, **d**). The NRP differentiate and express markers for both glutamatergic (glutamate, **b**) and GABAergic (GAD65/67, **e**) neurons. Scale bars = 50 μm

2 Materials

2.1 Preparation of Neuroepithelial Cells (NEP) from E10.5 Rat Spinal Cord

2.1.1 Fibronectin-Coated Dishes

1. Bovine Serum fibronectin.
2. Tissue Culture Water.
3. Hanks Balance Salt Solution without calcium and magnesium.
4. Tissue culture flask or glass coverslips.
5. T-25 tissue culture flask, 35 mm × 10 mm tissue culture dish, or glass coverslips can be used for the culture of NEP (*see* **Note 1**).

2.1.2 Chick Embryo Extract (CEE)

1. Fertilized chicken eggs.
2. Eagle's Minimal Essential Medium.
3. 20 ml syringes.
4. 60 ml syringes.
5. 50 ml conical tubes.
6. Hyaluronidase 1 mg/ml solution.
7. 0.45 μm sterile syringe filter.
8. 0.22 μm sterile syringe filter.

2.1.3 NEP Medium

1. DMEM/F12.
2. Bovine Serum Albumin (BSA).
3. B-27 (50× stock solution).
4. Pen/Strep (100× stock solution).
5. L-Glutamine (100× stock solution).
6. Fungizone (1,000× stock solution).
7. N2 supplement (100× stock solution).
8. NEP Complete Medium.
9. NEP Basal Medium.
10. Chick Embryo Extract (CEE).
11. bFGF.

2.1.4 NEP Dissection (All Materials Must Be Sterile)

1. Anesthetic.
2. E10.5 timed-pregnant rat.
3. Scissors.
4. 100 mm × 20 mm cell culture dishes.
5. DMEM/F12.
6. Tungsten needles.
7. Collagenase I/dispase II solution.
8. NEP complete medium.
9. Pasteur pipette.
10. #5 forceps.
11. 0.05 % Trypsin/EDTA.
12. Soybean trypsin inhibitor.
13. Fibronectin-coated T-25 flask or equivalent.

2.1.5 Passaging NEP Cells

1. 3 T-25 flasks, 1T-75 flask, or equivalent.
2. HBSS without calcium or magnesium.
3. 0.05 % Trypsin/EDTA.
4. Soybean trypsin inhibitor.

2.1.6 Freezing NEP

1. NEP Basal.
2. DMSO.
3. CEE.
4. Cryovials.

2.1.7 Thawing NEP

1. NEP basal medium.
2. NEP complete medium.
3. Fibronectin-coated culture dishes, T-25 flask, T-75 flask, or equivalent.

2.2 Preparation of Neuronal Restricted Progenitors and Glial Restricted Progenitor from E13.5 Rat Spinal Cord

2.2.1 Preparation of Poly-L-Lysine and Laminin Coated Dishes for NRP/GRP Culture

1. Poly-L-Lysine (PLL).
2. Tissue Culture Water.
3. 0.22 µm sterile syringe filter.
4. Natural Mouse Laminin.
5. Sterile Phosphate Buffered Saline (PBS).

2.2.2 NRP/GRP Medium

1. DMEM/F12.
2. BSA.
3. B-27 (50× stock solution).
4. Pen/Strep (100× stock solution).
5. N2 supplement (100×).
6. NRP/GRP complete medium.
7. NRP basal medium.
8. bFGF.
9. NT-3.

2.2.3 NRP/GRP Dissection

All materials must be sterile.

1. Anesthetic agent.
2. 70% ethanol.
3. 100 mm × 20 mm cell culture dish.
4. DMEM/F12.
5. E13.5–E14.0 timed-pregnant rat.
6. Microscissors.
7. #5 fine forceps.
8. Collagenase I/dispase II solution.
9. 50 ml conical tube.

2.2.4 NRP/GRP Dissociation

1. 0.05 % Trypsin/EDTA.
2. Soybean trypsin inhibitor.
3. NRP/GRP basal medium.

2.2.5 NRP/GRP Culture

1. NRP/GRP complete medium.
2. Poly-L-Lysine and Laminin T-75 tissue culture flask, or equivalent.

2.2.6 Enrichment of NRP/GRP

1. Anti-A2B5 IgM antibody (DSHB).
2. Anti-ENCAM IgM antibody (DSHB).

Table 1
A partial list of useful antibodies and dilutions

Antigen	Host species	Recommended dilution	Vendor	Product ID	Use
Nestin	Mouse IgG1	1:1,000	BD Pharmigen	556309	NEP, NRP/GRP
Sox2	Mouse IgG2A	1:200	R&D	MAB2018	NEP
A2B5	Mouse IgM	1:2	DSHB		GRP
ENCAM	Mouse IgM	1:2	DSHB		NRP
BIII Tubulin	Mouse IgG2a	1:500	Covance	MMS-435P	Neurons
BIII Tubulin	Rabbit	1:1,000	Covance	Prb-435P	Neurons
GFAP	Mouse IgG1	1:1,000	Millipore	MAB3402	Astrocytes
GFAP	Rabbit	1:2,000	Millipore	AB5804	Astrocytes
RIP	MouseIgG1	1:10,000	Millipore	MAB1580	Oligodendrocytes
O4	Mouse IgGM	1:50	R&D	MAB1326	Oligodendrocytes

Developmental Studies Hybridoma Bank (DSHB)

3. Goat anti-Mouse IgM, unlabeled (Millipore, AP128).
4. Sterile, bacteriological polystyrene petri dishes.
5. Freshly dissected NRP/GRP.

2.2.7 Freezing NRP/GRP

1. NRP/GRP Freezing Medium.
2. NRP/GRP Basal.
3. DMSO.
4. CEE.
5. NT-3.
6. bFGF.

2.3 Characterizing NEP and NRP/GRP Cultures In Vitro with Immunocytochemistry

Antibodies and dilutions

3 Methods

3.1 Preparation of Neuroepithelial Cells (NEP) from E10.5 Rat Spinal Cord

NEP are a multipotent cell type capable of producing neurons, astrocytes, and oligodendrocytes and NEP can self-renew in vivo. Therefore, NEP are a true NSC population and represent an intermediate step between less mature cells, such as ESC, and mature neural phenotypes. NEP are grown in an adherent culture on fibronectin-coated dishes rather than as neurospheres, which helps to maintain the uniformity of the culture and provides the opportunity for accurate observation of cultures with simple phase microscopy. NEP identity can be confirmed using antibodies for Nestin and Sox2. Isolation of NEP from the neural tubes of rats requires a combination of physical and chemical dissection techniques and should take 1–2 h per litter. We recommend having at least two technicians prepared to conduct the dissection, as rodent litters may vary between 2 and 15 embryos. Also take this into account when preparing medium, culture dishes, etc. (*see* **Note 2**).

3.1.1 Fibronectin-Coated Dishes

NEP can be expanded and maintained in their multipotent state on fibronectin-coated dishes. The following protocol describes how to use bovine fibronectin for coating. After coating, fibronectin-coated dishes are stable for 2 months at 4 °C.

1. Dilute fibronectin in Tissue Culture Water at 20 μg/ml.
2. Add enough fibronectin solution to the culture dish to just cover the bottom of the dish.
3. Place the dish on a rocker at 4 °C overnight.
4. Following day rinse flasks 2× with HBSS without calcium and magnesium.
5. Cover bottom of flask with HBSS for storage.
6. Coated dishes can be stored for up to 2 months.
7. Rinse flasks with warm (37 °C) tissue culture medium before use.

3.1.2 Collagenase I/Dispase II Solution

A solution of collagenase I/dispase II will be used to dissociate the neural tubes during the NEP dissection and may be used during the NRP/GRP dissection. It should always be made fresh the day of the dissection.

1. Combine 1 mg/ml collagenase I (Worthington) and 2 mg/ml Dispase II (Roche) in Hanks Balanced Salt Solution (HBSS) without calcium and magnesium.
2. Pass through a sterile 0.22 μm filter.

3.1.3 Chick Embryo Extract (CEE)

CEE is a critical component of the NEP medium and freezing medium for both NEP and NRP/GRP. CEE can be prepared well in advance and stored at −80°. In our experience the effectiveness of CEE varies by vendor. We recommend using the CEE from US Biological (San Antonio, TX, USA); otherwise follow the steps below to produce CEE.

1. 36–40 fertilized chicken eggs are incubated for 11 days at 37.5 °C.
2. Wash eggs with 70 % ethanol.
3. Remove embryos in small batches, decapitate with sterile forceps and rinse 3× in petri dish with ice-cold sterile Eagle's

Table 2
NEP Media

NEP Basal Medium	
Component	**Concentration**
DMEM/F12	94.9 %
Bovine Serum Albumin (BSA)	1 mg/ml
B-27 (50× stock solution)	2 %
Pen/Strep (100× stock solution)	1 %
L-glut (100× stock solution)	1 %
Fungizone (1,000× stock solution)	0.1 %
Pass medium through 0.22 μm sterile filter	
N2 supplement (100× stock solution)	1 %
NEP Complete Medium	
Component	**Concentration**
NEP Basal Medium	90 %
Chick Embryo Extract (CEE)	10 %
bFGF	20 ng/ml

Minimal Essential Medium (MEM) until all embryos are harvested.

4. Embryos are macerated by passage through a 20 ml syringe into 50 ml conical tubes.
5. Ten embryos produce about 25 ml of homogenate; add equal volume of MEM.
6. Rock the tubes at 4 °C for 1 h.
7. Add sterile hyaluronidase (1 mg/25 ml of chick embryo).
8. Centrifuge for 6 h at 30,000 × *g*.
9. Collect the supernatant into a 60 ml syringe and pass through a 0.45 μm filter.
10. Collect the filtrate into another 60 ml syringe and pass through a 0.22 μm filter.
11. Aliquot and store at −80 °C.

3.1.4 NEP Medium

NEP medium should be prepared under sterile conditions in a tissue culture hood to prevent contamination. The materials required are listed in Table 2 and should be prepared in DMEM/F12 for growth of cells in 5–10 % CO_2 environment. NEP basal medium can be used for manipulating cells during the dissection and other procedures, but NEP complete medium (NEP basal + bFGF and CEE, *see* Table 2) should be used for culturing cells. Complete medium should only be prepared as needed to save costs.

3.1.5 NEP Dissection

1. Anesthetize the dam with an appropriate agent, such as Euthasol.
2. When the dam is anesthetized, swab the abdomen with 70 % ethanol for 10 s.
3. Perform a laparotomy and remove both uterine horns.
4. Euthanize the dam with a thoracotomy.
5. Remove a sac from the uterine horn and rinse 1× in 70 % ethanol 2× in DMEM/F12.
6. Place in the lid of a sterile 100 mm × 20 mm cell culture dish partially filled with DMEM/F12 (*see* **Note 3**).
7. If the embryo is the correct gestational age it should be "C" shaped and it should be expelled from the sac when sac is cut. Count somites to insure correct gestational age (13–20 somites).
8. Dissect trunk segments of neural tube with surrounding somites and tissue (last ten somites) using tungsten needles.
9. Trunk segments are then incubated for approximately 7 min at room temperature with 1–2 ml of collagenase I/dispase II (*see* **Note 4**).
10. After the neural tubes begin to separate, carefully aspirate collagenase I/dispase II and replace with 2 ml NEP complete medium (*see* **Note 5**).
11. Carefully triturate with a Pasteur pipette to release neural tubes from the somites and connective tissue.
12. Remove any remaining somites with #5 forceps and/ or tungsten needles.
13. Remove the supernatant and incubate the neural tubes in 2 ml 0.05 % Trypsin/EDTA at 37 °C for 5 min.
14. After 5 min add an equal volume (2 ml) of soybean trypsin inhibitor.
15. Triturate the cells to dissociate the neural tubes.
16. Spin at 150–300 × *g* for 5 min.
17. Resuspend cells in 5 ml NEP complete medium and plate on a fibronectin-coated T-25 flask.
18. Incubate cells at 37 °C and 5 % CO_2.

3.1.6 Passaging NEP Cells

1. Grow cells to 80–90 % confluency before splitting.
2. Plan to split cells into three times the surface area (i.e., 1T-25 flask to 3T-25 flasks or 1T-75 flask).

Table 3
NEP freezing medium

Component	Concentration
NEP Basal	80 %
DMSO	10 %
CEE	10 %
Pass medium through 0.22 μm sterile filter	

3. Remove medium and rinse 2× with warmed HBSS.
4. Add 0.05 % Trypsin/EDTA (enough to cover base of culture dish).
5. When cells have lifted add equal volume of soybean trypsin inhibitor.
6. Add 10 ml of complete medium and spin at 150–300 × *g* for 5 min.
7. Decant supernatant and resuspend pellet in complete medium and plate on fibronectin-coated dishes.

3.1.7 Differentiation of NEP to NRP/GRP

1. Passage cells as described in Subheading 3.1.6 and plate on Poly-L-Lysine (PLL) and Laminin (LN) substrates (same substrate as NRP/GRP).
2. Remove CEE from culture medium to allow NEP to differentiate.
3. Add bFGF (30 ng/ml) to NEP complete medium to bias cells towards GRP.
4. Add bFGF (10 ng/ml) and NT-3 (10 ng/ml) to NEP complete medium to bias cells towards NRP.

3.1.8 Freezing NEP (See Table 3)

1. Cells should be approximately 90 % confluent.
2. Remove medium.
3. Rinse 2× with warmed HBSS to remove any dead or floating cells.
4. Incubate at 37 °C with 0.05 % Trypsin/EDTA.
5. Once cells have lifted, add an equal amount of soybean trypsin inhibitor.
6. Decant cells into a conical vial.
7. After cells are decanted, wash the flask 2× with 5 ml NEP complete medium to resuspend any cells that were left behind in the previous step. Decant and add to the same conical vial as the cell suspension.
8. Spin at 150–300 × *g* for 5 min.

9. Decant supernatant and resuspend cells in freezing medium (3 ml per confluent T-75 flask).
10. Aliquot 1 ml each into 2 ml cryovials.
11. Place cells in −80 °C overnight.
12. Transfer to liquid nitrogen following day.

3.1.9 Thawing NEP

1. Prior to thawing cells, warm (37 °C) 10 ml of NEP basal medium and 15 ml of NEP complete medium.
2. Take cryovial from liquid nitrogen and bury it in bucket of wet ice.
3. Thaw aliquot in 37 °C water bath until cells just begin to thaw.
4. Add 1 ml warmed medium drop by drop to cryovial.
5. Add the 2 ml suspension drop by drop to 10 ml warmed medium.
6. Rinse cryovial once with 1 ml warmed medium.
7. Spin at 150–300 × *g* for 5 min.
8. Decant supernatant.
9. Resuspend cells in 15 ml NEP complete medium and plate in a fibronectin-coated T-25 flask.

3.2 Preparation of Neuronal Restricted Progenitors and Glial Restricted Progenitor from E13.5 Rat Spinal Cord

NRP/GRP are prepared as a mixed culture of cells that are collectively capable of producing neurons, astrocytes, and oligodendrocytes, but NRP will only produce neurons whereas GRP will only produce astrocytes and oligodendrocytes. Therefore, NRP/GRP are not a true NSC population but represent a necessary intermediate step between multipotent NSC and mature neural or glial phenotypes. NRP/GRP are grown in an adherent culture on dishes coated with Poly-L-Lysine and Laminin. NRP can be identified by expression of embryonic neural cell adhesion molecule (E-NCAM) and GRP can be identified by expression of A2B5. Preparation of NRP/GRP from E13.5 embryos requires a primary physical dissection followed by chemical and mechanical dissociation and culturing. The full dissection should take 1–2 h and, as with NEP dissection, we recommend having at least 2 technicians prepared to conduct the dissection, as rodent litters may vary between 2 and 15 embryos.

3.2.1 Preparation of Poly-L-Lysine and Laminin Coated Dishes for NRP/GRP Culture

Preparation of Poly-L-Lysine/Laminin (PLL/LN) coated dishes is a two step process requiring 2 days to complete. The coating should be carried out at 4 °C on a rocker. Coated dishes are stable for 4 weeks after coating.

1. Prepare 10× (150 μg/ml) stock of PLL in Tissue Culture Water, freeze, and store at −20 °C.

2. PLL (Day 1):
   (a) Dilute stock solution of PLL with Tissue Culture Water to 15 μg/ml.
   (b) Filter PLL mixture with 0.22 μm syringe filter and 60 ml syringe.
   (c) Coat bottom of T-75 flask with 15 ml of 15 μg/ml PLL solution.
   (d) Place overnight on rocker at 4 °C.
3. Laminin (Day 2)
   (a) Laminin stocks must be kept at −80 °C and should be thawed for 2–4 h at 4 °C (*see* **Note 6**).
   (b) Aspirate PLL and wash 2× with PBS prior to adding Laminin.
   (c) Prepare 15 μg/ml solution of mouse Laminin in PBS. Do Not Filter.
   (d) Add 15 ml of 15 μg/ml Laminin in PBS to T-75 flask and coat on rocker overnight at 4 °C.
   (e) On third day, rinse 1× with PBS.
   (f) Add 15 ml of HBSS and store at 4 °C for up to 1 month.

3.2.2 NRP/GRP Medium

NRP/GRP medium should be prepared under sterile conditions in a tissue culture hood to prevent contamination. The materials required are listed in Table 4 and should be prepared in DMEM/F12 for growth of cells in 5–10 % CO_2 environment. NRP/GRP basal medium can be used for manipulating cells during the dissection and other procedures, but NRP/GRP complete medium (NRP/GRP basal + bFGF and NT-3, *see* Table 4) should be used for culturing cells.

3.2.3 NRP/GRP Dissection

1. Anesthetize the dam with an appropriate compound, such as Euthasol.
2. When the dam is anesthetized, swab the abdomen with 70 % ethanol for 10 s.
3. Perform a laparotomy and remove both uterine horns.
4. Euthanize the dam with a thoracotomy.
5. Place the entire uterus in a sterile 100 mm × 20 mm cell culture dish with DMEM/F12 on ice.
6. Remove a sac from the uterine horn and rinse the sac 1× in 70 % ethanol and 2× in DMEM/F12.
7. Place in the lid of a sterile 100 mm × 20 mm cell culture dish partially filled with DMEM/F12 (*see* **Note 7**).
8. Using microscissors, open the sac and release the embryo.

Table 4
NRP/GRP media

NRP/GRP basal medium	
Component	**Concentration**
DMEM/F12	96 %
BSA	1 mg/ml
B-27 (50× stock solution)	2 %
Pen/Strep (100× stock solution)	1 %
Pass medium through 0.22 μm sterile filter	
N2 supplement (100×)	1 %
NRP/GRP complete medium	
Component	**Concentration**
NRP basal medium	100 %
bFGF	20 ng/ml
NT-3	10 ng/ml

9. Measure the crown to rump distance to determine the embryonic age, the embryo should be between 8.5 and 10 mm long for an E13.5–E14 spinal cord.
10. Using a pair of microscissors, remove the head at the cisterna magna (*see* **Notes 8** and **9**).
11. Place the embryo with the ventral surface down.
12. Incise the midline skin over the neural tube (i.e., from the dorsal surface) using a pair of microscissors.
13. Firmly hold the embryo down with jeweler's forceps. Place the forceps in an area of the embryo that is not needed as it will be damaged (*see* **Note 10**).
14. Using sharp #4 or #5 forceps, dissect the spinal cord with associated meninges and DRGs from the embryo.
15. Peel off the meninges and DRGs as far caudally as possible. Place any cleaned sections of the spinal cord in a 50 ml conical tube of ice-cold DMEM/F12.
16. The lumbar cord may require the use of collagenase I/dispase II solution to remove meninges; if so place the lumbar cord in collagenase I/dispase II for 9 min (*see* **Note 11**).
17. Remove cord from collagenase I/dispase II and place in 100 mm × 20 mm dish with DMEM/F12.

18. Finish removing the meninges and place the cleaned spinal cord into 50 ml conical tube with DMEM/F12. Place up to three cords in one 50 ml conical tube.

3.2.4 NRP/GRP Dissociation

1. After the dissection is complete, centrifuge pooled cords at 150–300 × *g* for 10 min
2. Decant the supernatant from the 50 ml conical vial containing the cords.
3. Add 1 ml of 0.05 % trypsin–EDTA and incubate for at 37° for 10–20 min.
4. Using a sterile plugged 1,000 μl pipet tip, set at 800 μl, gently triturate cells about ten times until chunks of tissue look dissociated by eye.
5. Add equal volume 1 mg/ml soybean trypsin inhibitor to quench trypsin, and gently triturate with 10 ml pipet.
6. Do not over-triturate or many cells will die.
7. Pass cells suspension through a 40 μm filter- Cell strainer (BD Falcon 352340) into a fresh 50 ml tube.
8. Rinse the filter with another 10 ml NRP Basal Medium.
9. Centrifuge cell suspension at 150–300 × *g* for 5–10 min to get rid of the trypsin (*see* **Notes 12** and **13**).
10. Carefully decant the supernatant and add 1 ml of NRP basal medium.
11. Dilute 10 μl cells suspension 1:1 in Trypan blue and count with hemocytometer.
12. Discard the supernatant (*see* **Note 14**).

3.2.5 NRP/GRP Culture

1. Resuspend pellet with NRP/GRP complete medium (NRP/GRP basal medium, supplemented with 20 ng/ml bFGF and 10 ng/ml NT3) in PLL/LN-coated flask.
2. Recommended density: $1.5–3 \times 10^6$ cells/T-75 flask, or 5×10^5 to 1×10^6 cells/T-25 flask.
3. Incubate cells at 37 °C and 5 % CO_2.
4. NRP/GRP complete medium should be replaced every other day.

3.2.6 Differentiation of NRP/GRP Towards Neuronal Lineage

1. Maintain NRP/GRP on Poly-L-Lysine and Laminin substrate.
   (a) Remove bFGF.
   (b) Add retinoic acid (1 μM) to promote neuronal differentiation and NT-3 (10 ng/ml) to promote neuronal survival.

   OR

2. Maintain NRP/GRP on Poly-L-Lysine and Laminin substrate.
   (a) Remove bFGF.
   (b) Add 0.5 % High concentration Matrigel (Fig. 2).

3.2.7 GRP Enrichment and Differentiation Toward Glial Lineage

1. Maintain NRP/GRP on Poly-L-Lysine (100 μg/ml) substrate.
2. Remove NT-3.
3. Passage cells for 7–10 days prior to freezing to enrich for GRP.
4. Thaw enriched GRP on Poly-L-Lysine substrate (Fig. 3a, d).
5. Remove bFGF and NT-3.
6. Add BMP-4 (10 μg/ml; Fig. 3b, e) OR CNTF (10ug/ml; Fig. 3c, f).

3.2.8 Enrichment of NRP/GRP

NRP/GRP are isolated as a mixed culture, but some procedures may require enriched NRP or GRP. Pure NRP cultures are difficult to produce, especially in large numbers because NRP are generally more sensitive to manipulation than GRP. Despite these difficulties, enrichment for NRP is possible. Nearly pure populations of GRP can be produced reliably and with high quantities of cells by exploiting the greater ability of GRP to self-renew [23]. The difficulties in sorting NRP and GRP are due, in part, to the limitations of A2B5 and ENCAM IgM primary antibodies. IgM antibodies are, generally speaking, less suited to immunopanning and fluorescent activated cell sorting (FACS) than IgG antibodies.

Immunopanning is a procedure that uses an antibody covered substrate to bind a subset of cells to the substrate. Immunopanning is often conducted in a 2-phase procedure where cells are first negatively sorted (unwanted cells are bound to the substrate and discarded) followed by a positive selection (desired cells are placed on a new substrate, bound and collected). Thus, when sorting NRP/GRP both A2B5 and ENCAM substrates will always be prepared, only the order will be changed depending on the cell population that is desired.

1. *Step 1: Substrate Preparation*: Start of day1: Incubate two bacteriological petri dishes overnight at 4 °C with 10 ml of goat anti-mouse IgM (10 μg/ml) diluted in Tris–HCl 50 mM ph 9.5 (*see* **Note 15**).
2. Start of day 2: Remove the goat anti-mouse antibody.
3. Rinse the dish with HBSS without magnesium and calcium 3×.
4. Coat 1 dish with mouse-anti-ENCAM IgM hybridoma diluted 1:1 with HBSS without magnesium and calcium.

5. Coat the other dish with mouse-anti-A2B5 IgM hybridoma diluted 1:1 with HBSS without magnesium and calcium.
6. 2 h later remove the panning antibody solution.
7. Rinse the dishes 3× with HBSS without magnesium and calcium.
8. Store coated dishes with HBSS until ready for use.
9. Proceed to the negative selection.
10. *Step 2: Negative selection*: Begin with freshly isolated NRP/GRP suspended in 10 ml NRP/GRP complete medium (from Subheading 3.2.4).
11. The negative selection will remove undesirable cell types. Plate cells on dish coated with antibody specific to the undesirable cell type (i.e., if purified NRP are desired, plate the mixed NRP/GRP on the A2B5 coated dish).
12. Gently swirl the cells to insure the cells and evenly dispersed and there are no clumps.
13. Incubate the cells at *room temperature* for 1 h (*see* **Note 16**).
14. Gently tap the petri dish to remove any nonspecifically bound cells.
15. Remove the cell suspension and place in a 15 ml conical tube.
16. Discard the bound fraction.
17. Spin the cell suspension at 150–300 ×*g* for 5 min.
18. Resuspend the cells in 10 ml NRP/GRP complete medium.
19. *Step 3: Positive selection*: The positive selection will bind the desirable cell type to the plate. Plate cells on dish coated with antibody specific to the desirable cell type (i.e., if purified NRP are desired, plate the mixed NRP/GRP on the ENCAM coated dish).
20. Add cell suspension to ENCAM coated dish.
21. Gently swirl the cells to insure the cells and evenly dispersed and there are no clumps.
22. Incubate the cells at *room temperature* for 1 h (*see* **Note 16**).
23. Gently tap the petri dish to remove any nonspecifically bound cells.
24. Remove the cell suspension and discard.
25. Wash the ENCAM plate 2× with NRP/GRP basal medium to remove any nonspecifically bound cells.
26. Use a cell scraper to remove the bound cells.
27. Resuspend the cells in 10 ml NRP/GRP complete medium.
28. Plate the cells on Poly-L-Lysine/Laminin coated dishes (or other desired substrate).

Table 5
NRP/GRP freezing media

NRP/GRP freezing medium	
Component	**Concentration**
NRP/GRP Basal	80 %
DMSO	10 %
CEE	10 %
NT-3	20 ng/ml
bFGF	30 ng/ml
Pass medium through 0.22 μm sterile filter	

3.2.9 Freezing NRP/GRP (See Table 5)

1. Cells should be approximately 80 % confluent.
2. Remove medium.
3. Rinse 2× with warmed HBSS to remove any dead or floating cells.
4. Incubate at 37 °C with 0.05 % Trypsin–EDTA.
5. Once cells have lifted, add an equal amount of soybean trypsin inhibitor.
6. Transfer cells into a conical vial.
7. Add 5–10 ml of complete medium, count cells.
8. Spin at 150–300 ×*g* for 5 min.
9. Decant supernatant and resuspend cells in freezing medium (3 ml per confluent T-75 flask) (*see* **Note 17**).
10. Aliquot 1 ml each into 2 ml cryovials.
11. Place cells in −80 °C overnight (*see* **Note 18**).
12. Transfer to liquid nitrogen following day.

3.2.10 Thawing NRP/ GRP

1. Prior to thawing cells, warm (37 °C) 10 ml of NRP basal medium and 15 ml of NRP complete medium.
2. Take cryovial from liquid nitrogen and bury it in bucket of wet ice.
3. Thaw aliquot in 37 °C water bath until cells just begin to thaw.
4. Add 1 ml warmed medium drop by drop to cryovial.
5. Add the 2 ml suspension drop by drop to 10 ml warmed medium.
6. Rinse cryovial once with 1 ml warmed medium.
7. Spin at 150–300 ×*g* for 5 min.

8. Decant supernatant.
9. Resuspend cells in 15 ml NRP/GRP complete medium and plate in a PLL/LM coated T-75 tissue culture flask.

3.3 Characterizing NEP and NRP/GRP Cultures In Vitro with Immunocytochemistry

NEP and NRP/GRP can be easily identified in culture with immunocytochemistry. NEP express Nestin and Sox2, NRP express Nestin and ENCAM, and GRP express Nestin and A2B5. E-CAM and A2B5 are cell surface antigens and should be performed as a live stain. Other markers, such as those for mature neurons (BIII tubulin), astrocytes (GFAP), and oligodendrocytes (O4), can be performed on fixed cells.

3.3.1 Staining Live Cells

1. Decant medium.
2. Wash with 37 °C HBSS 2×.
3. Add 37 °C complete medium with primary antibody (Table 1).
4. Incubate at 37 °C and 5 % CO_2 for 30 min.
5. Decant medium.
6. Wash with warm HBSS 2×.
7. Add warmed complete medium with secondary antibody (1:400).
8. Incubate at 37 °C and 5 % CO_2 for 30 min.
9. Wash with 37 °C HBSS 2×.
10. Add 4 % PFA for 10 min.
11. Wash 3× with PBS.
    (a) Option 1: Counterstain with DAPI (1:1,000), coat with aqueous cover slipping medium and add coverslip.
    (b) Option 2: Store in PBS at 4 °C for additional staining at a later time.
    (c) Option 3: Perform another stain as described in the following section.

3.3.2 Staining Fixed Cells

1. Decant medium.
2. Add 4 % Paraformaldehyde (PFA) for 10 min.
3. Wash with PBS for 5 min 3×.
4. Block for 30 min in 5 % milk or 5 % normal serum at room temperature.
5. Treat with 0.2 % Triton in PBS if needed.
6. Add primary antibody (Table 1) in PBS with 2 % milk or serum.
7. Incubate for 30 min at room temperature.
8. Wash with PBS for 5 min 3×.
9. Add secondary antibody in PBS with 2 % milk or serum (*see* **Note 19**).

10. Incubate for 30 min at room temperature.
11. Wash with PBS for 5 min 3×.
12. Counterstain with DAPI (1:1,000).
13. Coat with aqueous cover slipping medium and add coverslip.

3.4 Summary

Neural stem cells and neural progenitor cells can be a useful source of neurons for in vitro and in vivo studies of neural function. Both NSC, such as neuroepithelial cells (NEP), and NPC, such as neuronal and glial restricted progenitors (NRP/GRP), can be directly isolated from the developing mammalian spinal cord or from pluripotent sources. The techniques discussed here should provide the ability to dissociate culture, freeze, thaw, and differentiate NEP or NRP/GRP towards mature neuronal and glial phenotypes.

4 Notes

1. Smaller culture dishes are recommended due to the small yield from NEP dissection.
2. Dams often produce fewer pups as a result of the first pregnancy. If using an in-house breeding colony using the second or third pregnancy may produce more pups per litter.
3. Using DMEM/F12 without phenol red may make the neural tubes easier to visualize.
4. Inspect the neural tubes under a dissecting scope and stop the collagenase1/dispase II reaction when there is evidence that the tubes and connective tissue are separating.
5. Alternatively, tubes can be moved with a pipette to a fresh vial of NRP complete medium.
6. If using small amounts of Laminin for coating coverslips, Laminin can be thawed, aliquoted, and refrozen. We do not recommend multiple freeze–thaw cycles.
7. Using the lid rather than the base provides better angles for the dissection because the lid sidewalls are lower.
8. Including more rostral sections of the embryo will produce cultures with serotonergic neurons from the raphe nucleus.
9. Other protocols recommend removing the head as the last step and instead using the head to hold the embryo during the dissection.
10. If the embryo will not remain on the ventral surface, ensure that the head has been completely removed and remove the tail if necessary.
11. To prevent overtreatment with collagenase I/dispase II, monitor the cords every few minutes for signs that the meninges are separating.

12. Usually 1,000 rpm for bench top centrifuges capable of accepting 50 ml conical tubes, but consult your centrifuge manufacturer for specific settings.
13. Lower speeds and longer times may be used if desired.
14. Beginners may wish to decant the supernatant to another conical tube. If cell yields are lower than expected, some cells may be found in the supernatant. This would indicate that the pellet was disturbed after centrifugation and should be handled more cautiously.
15. Using tissue culture dishes will promote nonspecific adhesion during the selection steps.
16. Try not to move the culture dish during the incubation period.
17. A2B5 and ENCAM are IgM antibodies and require anti-IgM specific secondary antibodies.
18. Using sterile DMSO can eliminate the need to filter freezing medium.
19. Specially designed cooling containers can improve cell survival in the freezing stage (e.g., Nalgene #5100-0001).

References

1. See J, Bonner JF, Neuhuber B, Fischer I (2010) Neurite outgrowth of neural progenitors in presence of inhibitory proteoglycans. J Neurotrauma 27:951–957
2. Ketschek AR, Haas C, Gallo G, Fischer I (2012) The roles of neuronal and glial precursors in overcoming chondroitin sulfate proteoglycan inhibition. Exp Neurol 235:627–637
3. Dimos JT, Rodolfa KT, Niakan KK et al (2008) Induced pluripotent stem cells generated from patients with ALS can be differentiated into motor neurons. Science 321: 1218–1221
4. Egawa N, Kitaoka S, Tsukita K et al (2012) Drug screening for ALS using patient-specific induced pluripotent stem cells. Sci Transl Med 4:145ra104
5. Kim J, Efe JA, Zhu S, Talantova M, Yuan X, Wang S, Lipton SA, Zhang K, Ding S (2011) Direct reprogramming of mouse fibroblasts to neural progenitors. Proc Natl Acad Sci U S A 108:7838–7843
6. Mujtaba T, Piper DR, Kalyani A, Groves AK, Lucero MT, Rao MS (1999) Lineage-restricted neural precursors can be isolated from both the mouse neural tube and cultured ES cells. Dev Biol 214:113–127
7. Mujtaba T, Mayer-Proschel M, Rao MS (1998) A common neural progenitor for the CNS and PNS. Dev Biol 200:1–15
8. Cao Q-L, Howard RM, Dennison JB, Whittemore SR (2002) Differentiation of engrafted neuronal-restricted precursor cells is inhibited in the traumatically injured spinal cord. Exp Neurol 177:349–359
9. Han SS, Kang DY, Mujtaba T, Rao MS, Fischer I (2002) Grafted lineage-restricted precursors differentiate exclusively into neurons in the adult spinal cord. Exp Neurol 177:360–375
10. Lepore AC, Han SS, Tyler-Polsz CJ, Cai J, Rao MS, Fischer I (2004) Differential fate of multipotent and lineage-restricted neural precursors following transplantation into the adult CNS. Neuron Glia Biol 1:113–126
11. Theele DP, Schrimsher GW, Reier PJ (1996) Comparison of the growth and fate of fetal spinal iso- and allografts in the adult rat injured spinal cord. Exp Neurol 142:128–143
12. Wirth ED 3rd, Reier PJ, Fessler RG, Thompson FJ, Uthman B, Behrman A, Beard J, Vierck CJ, Anderson DK (2001) Feasibility and safety of neural tissue transplantation in patients with syringomyelia. J Neurotrauma 18:911–929
13. White TE, Lane MA, Sandhu MS, O'Steen BE, Fuller DD, Reier PJ (2010) Neuronal progenitor transplantation and respiratory outcomes following upper cervical spinal cord injury in adult rats. Exp Neurol 225:231–236

14. Kalyani A, Hobson K, Rao MS (1997) Neuroepithelial stem cells from the embryonic spinal cord: isolation, characterization, and clonal analysis. Dev Biol 186:202–223
15. Shihabuddin LS, Horner PJ, Ray J, Gage FH (2000) Adult spinal cord stem cells generate neurons after transplantation in the adult dentate gyrus. J Neurosci 20:8727–8735
16. Cao QL, Zhang YP, Howard RM, Walters WM, Tsoulfas P, Whittemore SR (2001) Pluripotent stem cells engrafted into the normal or lesioned adult rat spinal cord are restricted to a glial lineage. Exp Neurol 167:48–58
17. Kalyani AJ, Piper D, Mujtaba T, Lucero MT, Rao MS (1998) Spinal cord neuronal precursors generate multiple neuronal phenotypes in culture. J Neurosci 18:7856–7868
18. Rao MS, Noble M, Mayer-Proschel M (1998) A tripotential glial precursor cell is present in the developing spinal cord. Proc Natl Acad Sci U S A 95:3996–4001
19. Yang H, Mujtaba T, Venkatraman G, Wu YY, Rao MS, Luskin MB (2000) Region-specific differentiation of neural tube-derived neuronal restricted progenitor cells after heterotopic transplantation. Proc Natl Acad Sci U S A 97: 13366–13371
20. Bonner JF, Blesch A, Neuhuber B, Fischer I (2010) Promoting directional axon growth from neural progenitors grafted into the injured spinal cord. J Neurosci Res 88: 1182–1192
21. Lepore AC, Bakshi A, Swanger SA, Rao MS, Fischer I (2005) Neural precursor cells can be delivered into the injured cervical spinal cord by intrathecal injection at the lumbar cord. Brain Res 1045:206–216
22. Bonner JF, Connors TM, Silverman WF, Kowalski DP, Lemay MA, Fischer I (2011) Grafted neural progenitors integrate and restore synaptic connectivity across the injured spinal cord. J Neurosci 31:4675–4686
23. Haas C, Neuhuber B, Yamagami T, Rao M, Fischer I (2012) Phenotypic analysis of astrocytes derived from glial restricted precursors and their impact on axon regeneration. Exp Neurol 233:717–732

Chapter 8

Derivation of Neuronal Cells from Fetal Normal Human Astrocytes (NHA)

S. Ausim Azizi and Barbara Krynska

Abstract

Pathological changes in most neurological diseases are marked by cell loss. To understand the mechanisms of neurogenesis and brain repair at a cellular level, observation on less complex systems provide valuable knowledge which offer the basis for therapeutic interventions. This has been the impetus for neural cell culture studies and the development of in vitro models. Here, we provide protocols for differentiation into neuronal lineage of commercially available normal human astrocytes (NHA) that are isolated from normal fetal human brain (Lonza, Inc.). It is known that some of GFAP positive astrocytic cells have stem/progenitor cell characteristics; however, understanding of the human GFAP positive cells with these characteristics remains limited. The genesis of neuronal lineage cells from the NHA occurs in adherent culture conditions by removal of serum and exposure to bFGF. When transferred to serum-free medium supplemented with bFGF, NHA cells generate neuronal precursors that express doublecortin, nestin and are negative for GFAP. After withdrawal of bFGF they mature into neurons. The average time required for generation of neuronal cells using this protocol is about 3 weeks. Our model of neurogenesis captures a contained in vitro system consisting of both neurons and glia. This "human brain in a dish" model can be used to assay the effects of interventions on developing human neurons at a cellular and molecular level and is also suitable for modeling of various aspects of human diseases and testing of novel therapeutic strategies.

Key words Neurogenesis, Astrocytes, Neurons, Neuronal lineage, Differentiation, Cell culture

1 Introduction

Understanding the mechanisms of neurogenesis is of interest for scientific and therapeutic purposes. Classical models of neurogenesis are based on the notion that separate committed neuronal and glial precursors originate from a common nestin-positive neural stem cell (NSC). During differentiation, these neural stem cells lose expression of nestin, and gain expression of more specialized proteins such as β-III tubulin or GFAP [1]. The traditional model of neural development, exhibiting that diverged neuronal and glial precursors emerge from neuroepithelial stem cells in the

Shohreh Amini and Martyn K. White (eds.), *Neuronal Cell Culture: Methods and Protocols*, Methods in Molecular Biology, vol. 1078, DOI 10.1007/978-1-62703-640-5_8,

embryonic ventricular zone, has been challenged after discoveries that radial glia can be neural progenitors, and can give rise to neurons in the developing brain [2, 3]. Neurogenesis has also been shown in the mature brain, and it is thought that some astrocytes in the subventricular zone (SVZ) and the subgranular zone (SGZ) maintain their neurogenic potential and generate neurons through life [4]. Populations of multipotent astrocytic cells, giving rise to both neurons and glia, have been recently isolated not only from the SVZ but also other regions of the immature mammalian brain [5]. However, our understanding of neurogenesis has remained limited owing to the deficiency of appropriate experimental systems. Here we describe a method for the initiation of neuronal precursors derived from human GFAP-expressing cells, designated NHA, and for further expansion/differentiation of NHA-derived precursors. During the initial proliferative phase, NHA cells co-express GFAP, nestin and low levels of β-III tubulin [6]. After our initial report, similar findings on GFAP and β-III tubulin co-expression in primary cultures of fetal astrocytes and in populations of progenitor-like cells of the ventricular/subventricular zones and the cortical plate were published by other laboratories [7]. In order to determine the generation of neuronal linage in NHA cultures, it is critical to carefully determine the identity of neuronal precursors to exclude the population of cells with overlapping marker expression such as β-III tubulin and GFAP positive cells. In addition to phenotypic characteristics, there are also evident morphological changes in NHA cells undergoing during differentiation of NHA-derived neuronal precursors that are easy to observe in an adherent cell culture system. Access to protocols for differentiation of human multipotent astrocytic cells into neuronal lineage provides a model system to study neurogenesis and the development of neurological diseases. We have recently shown the use of NHA cell proliferation/differentiation system for modeling of the differentiation potential of human glioma cells by comparing the differentiation capacity of glioma-derived cells and human NHA cells [8]. Direct comparison between differentiation of normal human multipotent glial cells and tumor cells isolated from human gliomas constitutes an important and valuable method for demonstrating their phenotypic characteristics, differentiation and better understanding of tumor biology. Besides studying oncogenesis, the NHA differentiation protocol presented here may serve as a valuable tool for studying developmental anomalies and other neurological diseases. Such a system may also be uniquely useful for studying the effects of drugs on developing human neurons and validation of novel treatment strategies. In addition, the protocol presented here can serve as a base for further studying the stem-like properties of NHA cells and refining of methods for differentiation of NHA cells.

2 Materials

2.1 Reagents

1. NHA cells, isolated from the cerebrum of 5 month old, human fetuses; commercially available at Lonza (Walkersville, MD; Clonetics™, NHA-Astrocytes AGM, Cryogenic ampule).
2. Astrocyte Growth Medium Lonza CC-3186 (Clonetics™, AGM BulletKit Medium).
3. 0.25 % Trypsin–EDTA (Life Technologies, Grand Island, NY; 25200-056).
4. Growth Factor Reduced Matrigel, Phenol Red-Free (BD Biosciences, San Jose, CA).
5. DMEM/F12 medium (Life Technologies, Grand Island, NY; 25200-056).
6. Anti-Laminin antibody (BD Biosciences, San Jose, CA; 354232).
7. Anti-GFAP antibody; glial fibrillary acidic protein (DAKOUSA, Carpinteria, CA; Z0334).
8. Anti-SOX2 antibody (EMD Millipore, Billerica, MA; AB5603).
9. Anti-Nestin antibody (BD Biosciences, San Jose, CA; 611659).
10. Anti-Doublecortin antibody (Santa Cruz Biotechnology, Santa Cruz, CA; SC-8066).
11. Anti-GABA antibody (Sigma, St. Louis, MO; A2052).
12. Anti-MAP2 antibody (EMD Millipore, Billerica, MA; MAB3418).

2.2 Reagent Preparation (See Note 1)

1. *For expansion of NHA cells.* Prepare the expansion medium (Astrocyte Growth Medium, AGM Bullet Kit, Lonza) according to the manufacturer's protocol. After growth supplements are added to basal medium, store ready to use medium at 4 °C for no more than 1 month.
2. *DMEM/F12 and bFGF medium.* Supplement DMEM/F12 with 10 ηg/ml of bFGF and N2 supplement immediately prior to use.
3. *Growth factor-free DMEM/F12 medium.* Supplement DMEM/F12 with N2 supplement and 0.2 mM ascorbic acid immediately prior to use.
4. *Matrigel.* Thaw the frozen vial of Matrigel overnight on ice at 4 °C. Prepare 1 ml aliquots in 1.5 ml Eppendorf tubes using chilled pipettes, tips and tubes and store at −20 °C (*see* **Note 2**).
5. *Matrigel coated chamber slides.* Coat chamber slides with 1 ml of diluted Matrigel per chamber at room temperature under the hood for 1 h. After 1 h, gently aspirate off the diluted Matrigel and rinse gently with room temperature serum-free medium once and aspirate. Slides are ready to use (*see* **Note 3**).

6. *Poly-D-Lysine coating*. Coat chamber slides with 1 ml of poly-D-lysine (0.1 mg/ml in sterile dH_2O) at room temperature for 10 min. Aspirate off the poly-D-lysine, rinse twice with 1 ml of sterile dH_2O, remove the dH_2O and allow to air dry for 30 min.
7. *Poly-D-Lysine and Laminin coating*. First coat chamber slides with poly-D-lysine as above and then coat chamber slides with 1 ml of Laminin (1 μg/ml in sterile dH_2O) overnight at room temperature. The following morning, aspirate off the Laminin and allow the plates to dry for 1 h prior to use.

3 Methods

3.1 Expansion of NHA Cells

1. Quickly thaw cryopreserved NHA cells immediately after receipt in a 37 °C water bath by gently swirling the vial.
2. Expand the NHA cells by plating the cells onto petri dishes (100 mm petri dish) at a density of 5,000 cells/cm^2 in 12 ml of expansion medium at 37 °C, 5 % CO_2 humidified incubator (*see* **Note 4**). Adjust all volumes accordingly for other size flasks. The next day after seeding, remove the expansion medium from the dish and replace it with the pre-warmed (37 °C), fresh medium (*see* **Note 5**). Thereafter, replace 50 % of the culture medium with the fresh medium every 2–3 days.
3. During the proliferation phase, parallel cultures of cells (plated on poly-D-lysine coated chamber slides) can be stained for neuronal, glial and stem cell markers to determine the phenotype of cultured cells. Proliferating NHA cells should co-express GFAP, SOX-2, nestin and low levels of β-III tubulin (Fig. 1 and ref. 6). Characterization of NHA cells should include morphological observation throughout the culture period.
4. When NHA cells are about 70–80 % confluent harvest the cells. Discard the expansion medium from the petri dish, rinse cells with 1× PBS, aspirate and add pre-warmed trypsin (3–4 ml per 100 mm petri dish (*see* **Note 6**)).

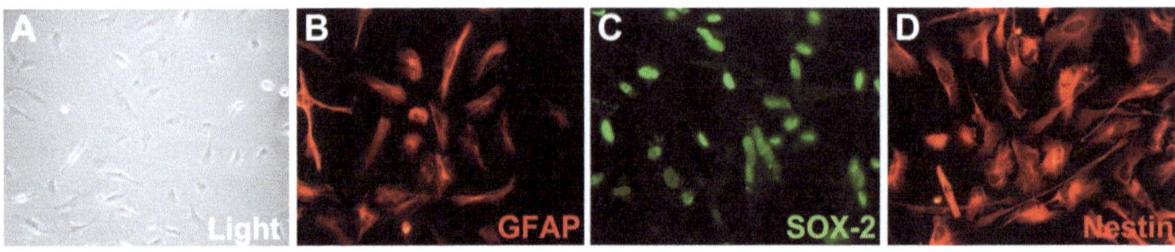

Fig. 1 Propagation of NHA cells. Morphology of NHA cells cultured in expansion medium at day 3 in culture (**a**), and expression of stem cell-related markers, GFAP (**b**), SOX-2 (**c**), and nestin (**d**). Original magnification: (**a**; ×20) and (**b–d**; ×40)

5. Incubate at 37 °C for 4–5 min and check under the microscope (*see* **Note 7**). Once the cells have detached, harvest the cells in DMEM/F12 medium supplemented with 10 % fetal bovine serum to inactivate the trypsin.
6. Pipette and triturate the cell suspension, remove 100 μL of cell suspension for counting with a hemocytometer using trypan blue to check for cell viability, and centrifuge the remaining cells.
7. To begin the differentiation of NHA cells immediately, separately centrifuge the number of cells required for proceeding with the differentiation experiment and the remaining cells designated for making freezing stocks.
8. Centrifuge the cell suspension for 7 min at 215 ×*g* and aspirate off the supernatant.
9. Pipette and triturate the pellet with freezing medium, and dispense cells into cryogenic vials making aliquots of 10^6 cells/1 ml of freezing medium labeled as passage one (P1) NHA cells, and store in liquid nitrogen for future experiments.
10. Pipette and triturate the cell pellet with differentiation medium adjusted based on the calculated seeding density, and proceed with the differentiation experiment.

3.2 Differentiation of NHA Cells

1. For differentiation (induction of neuronal precursors), plate P1 NHA cells from initial culture, or frozen stocks, at the density of 15,000 cells/cm^2 on Matrigel-coated petri dishes or chamber slides and culture overnight in expansion medium (*see* **Note 8**). Alternatively, poly-D-lysine and Laminin coated slides and dishes can be used instead of Matrigel.
2. The next day, gently aspirate the expansion medium using a sterile pipette and replace it with the pre-warmed DMEM/F12, supplemented with 10 ηg/ml of bFGF and N2 supplement, and culture cells for an additional 7–10 days.
3. Every 2 days replace 50 % of the culture medium with the fresh medium.
4. Neuronal precursors can be observed by day 7 of differentiation, based on cell morphology, by the appearance of adherent clusters of small cells on top of a layer of larger, flat cells and the phenotype by positive doublecortin and negative GFAP staining (Fig. 2). GFAP-negative and doublecortin positive neuronal precursors should be interspersed with larger GFAP positive cells. Neuronal precursors can be stained with other markers expressed by immature neurons, such as β-III tubulin and nestin (*see* **Note 9**).

5. For differentiation (maturation of neuronal precursors) on days 8–10 in culture, gently aspirate medium with bFGF using a sterile pipette and then replace it with the pre-warmed, growth factor-free DMEM/F12 medium supplemented with N2 and 0.2 mM ascorbic acid.
6. Allow the cultures to differentiate for an additional 8–10 days and replace 50 % of the culture medium with the fresh medium every 2–3 days (*see* **Note 10**).
7. The presence of cells differentiated along neuronal lineage should be determined based on cell morphology and expression of neuronal cell markers. These cells should be negative for GFAP, positive for β-III tubulin and MAP-2, and should readily differentiate toward GABA positive and a population of Tyrosine Hydroxylase (TH) - positive neuronal cells (Fig. 3). These cultures also contain GFAP - positive astrocytes (Fig. 3).

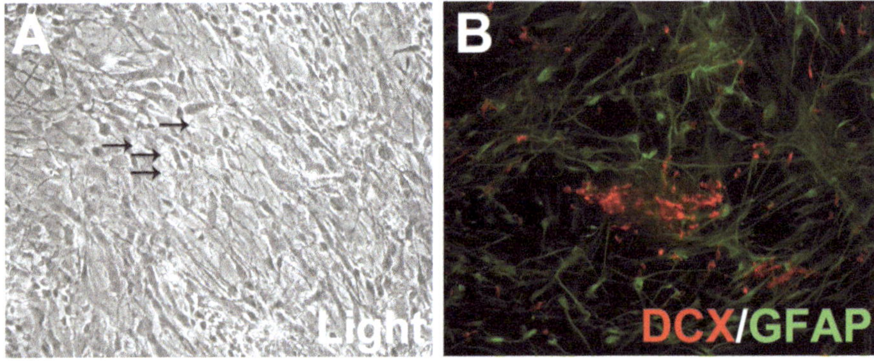

Fig. 2 Differentiation of NHA cells into neuronal lineage in serum-free medium supplemented with bFGF. At day 7 in culture there were detectable changes in cell morphology and phenotype. NHA cells generated clusters of proliferating neuronal precursors (**a**) that were positive for doublecortin (*red*) and negative for GFAP (*green*) (**b**). Original magnification: (×20)

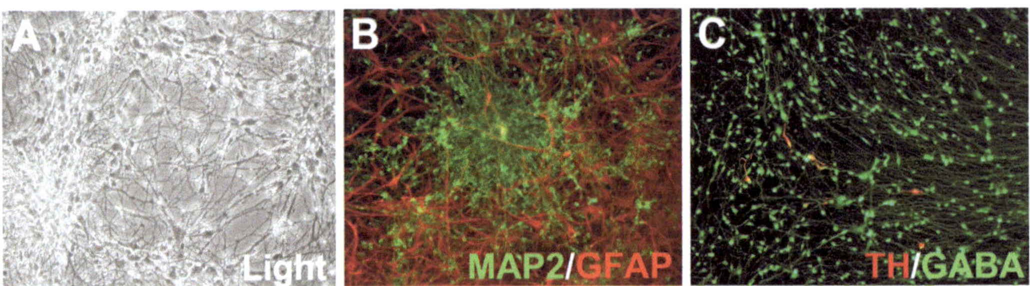

Fig. 3 Maturation of neuronal precursors in serum-free medium. Neuronal precursors further cultured in serum-free medium without growth factors differentiated into neuronal cells (*arrows*) (**a**) positive for MAP-2 (*green*) that were interspersed with GFAP positive astrocytes (*red*) (**b**). The majority of NHA cells differentiated into GABA-expressing cells and a population of Tyrosine Hydroxylase (TH) positive cells (**c**). Original magnification: (×20)

4 Notes

1. Sterilize all media using 0.22 μm filtration.
2. Matrigel must be thawed slowly to prevent gelatinization. Chilled pipettes, tips and tubes should be used when making aliquots of Matrigel.
3. Each lot of Matrigel has a different protein concentration. When diluting Matrigel do not dilute to less than 1 mg/ml of protein and dilute using serum-free medium that has been chilled on ice.
4. It is critical to culture NHA cells at the proper density as low seeding density causes loss of NHA differentiation capacity.
5. Avoid repeated warming and cooling of the medium. If the entire contents are not needed for a single procedure, transfer only the required volume to a sterile container and warm the medium to 37 °C.
6. To keep Trypsin–EDTA fresh and active after thawing, you may aliquot it into sterile tubes and re-freeze at −20 °C.
7. Examine the culture under the microscope and allow the trypsinization to continue until nearly all cells have detached.
8. In order to assure consistent differentiation for induction of neuronal precursors, use NHA cells after a single passage in expansion medium (P1). Using cells at higher passages may impact differentiation efficiency.
9. In order to determine the generation of neuronal precursors on day 6 or 7 of differentiation, parallel cultures can be monitored by immunocytochemistry. It is critical to carefully validate the identity of neuronal lineage to exclude any cells with overlapping marker expression such as GFAP and β-III tubulin positive cells.
10. If required at this stage, cultures can be maintained for an additional period of 5–7 days.

Acknowledgments

The authors wish to thank past and present members of the Department of Neurology and the Shriners Hospitals Pediatric Research Center/Center for Neural Repair and Rehabilitation for their continued support, insightful discussion, and sharing of ideas.

References

1. Okabe S, Forsberg-Nilsson K, Spiro AC, Segal M, McKay RD (1996) Development of neuronal precursor cells and functional postmitotic neurons from embryonic stem cells in vitro. Mech Dev 59:89–102
2. Alvarez-Buylla A, Garcia-Verdugo JM, Tramontin AD (2001) A unified hypothesis on the lineage of neural stem cells. Nat Rev Neurosci 2:287–293
3. Malatesta P, Hartfuss E, Gotz M (2000) Isolation of radial glial cells by fluorescent-activated cell sorting reveals a neuronal lineage. Development 127:5253–5263
4. Doetsch F (2003) The glial identity of neural stem cells. Nat Neurosci 6:1127–1134
5. Laywell ED, Rakic P, Kukekov VG, Holland EC, Steindler DA (2000) Identification of a multipotent astrocytic stem cell in the immature and adult mouse brain. Proc Natl Acad Sci U S A 97:13883–13888
6. Rieske P, Azizi SA, Augelli B, Gaughan J, Krynska B (2007) A population of human brain parenchymal cells express markers of glial, neuronal and early neural cells and differentiate into cells of neuronal and glial lineages. Eur J Neurosci 25:31–37
7. Draberova E et al (2008) Class III beta-tubulin is constitutively coexpressed with glial fibrillary acidic protein and nestin in midgestational human fetal astrocytes: implications for phenotypic identity. J Neuropathol Exp Neurol 67: 341–354
8. Witusik M et al (2008) Successful elimination of non-neural cells and unachievable elimination of glial cells by means of commonly used cell culture manipulations during differentiation of GFAP and SOX2 positive neural progenitors (NHA) to neuronal cells. BMC Biotechnol 8:56

Chapter 9

Slice Culture Modeling of Central Nervous System (CNS) Viral Infection

Kalen R. Dionne and Kenneth L. Tyler

Abstract

The complexity of the central nervous system (CNS) is not recapitulated in cell culture models. Thin slicing and subsequent culture of CNS tissue has become a valued means to study neuronal and glial biology within the context of the physiologically relevant tissue milieu. Modern membrane-interface slice culturing methodology allows straightforward access to both CNS tissue and feeding medium, enabling experimental manipulations and analyses that would otherwise be impossible in vivo. CNS slices can be successfully maintained in culture for up to several weeks for investigation of evolving pathology and long-term intervention in models of chronic neurologic disease.

Herein, membrane-interface slice culture models for studying viral encephalitis and myelitis are detailed, with emphasis on the use of these models for investigation of pathogenesis and evaluation of novel treatment strategies. We describe techniques to (1) generate brain and spinal cord slices from rodent donors, (2) virally infect slices, (3) assess virally induced injury/apoptosis, (4) characterize "CNS-specific" cytokine production, and (5) treat slices with cytokines/pharmaceuticals. Although our focus is on CNS viral infection, we anticipate that the described methods can be adapted to address a wide range of investigations within the fields of neuropathology, neuroimmunology, and neuropharmacology.

Key words Organotypic, Ex vivo, Brain slice, Spinal cord slice, Virus, Encephalitis, Myelitis, Cytokine, Apoptosis, Caspase

1 Introduction

In the mature brain, ~100 billion neurons interconnect to form $>10^{15}$ synapses [1]. There are another 100–5,000 billion glial cells (i.e., astrocytes, oligodendrocytes, microglia, and ependymal cells), which not only provide spatial organization to the nervous system, but also support neurological homeostasis, participate in signaling events, and mediate immunologic responses [2, 3]. Due to its inherent complexity, the organizational and functional aspects of the central nervous system (CNS) cannot be completely recapitulated in cell line and primary cell culture models. While such reductionist approaches are amenable to a wide range of experimental protocols, these systems often fail to reveal the fundamental nature

Shohreh Amini and Martyn K. White (eds.), *Neuronal Cell Culture: Methods and Protocols*, Methods in Molecular Biology, vol. 1078, DOI 10.1007/978-1-62703-640-5_9, © Springer Science+Business Media New York 2013

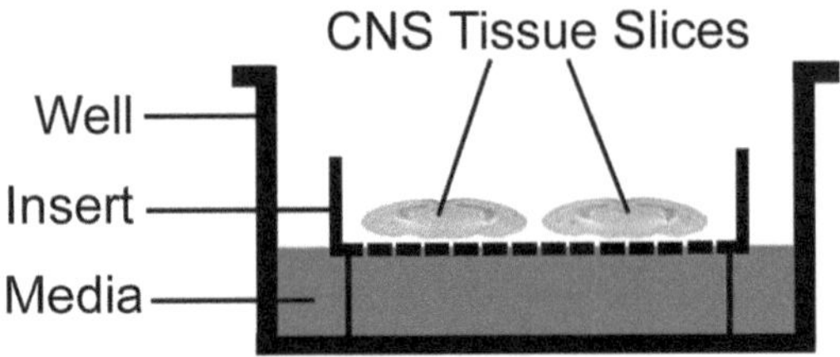

Fig. 1 Schematic of membrane-interface culturing system. Cutaway view of a single well depicts CNS tissue slices suspended at the interface of media and the incubator atmosphere. This configuration allows both nutrient absorption and gas exchange for long-term viability

of in vivo neurobiology. Although in vivo studies provide highly relevant information about systems biology, animal experimentation is procedurally challenging, time consuming, and—in the case of disease models—debatably inhumane.

Ex vivo culturing of CNS tissue is a uniquely powerful method that provides neuroscientists with in vitro-like experimental versatility and in vivo-like experimental validity. Culture of nervous tissue dates back to 1907, when neuronal outgrowth was observed from embryonic frog explants affixed to coverslips [4]. Successful long-term culture of explanted mammalian CNS tissue, however, was first achieved 40 years later with the introduction of the Roller-tube method [5–8], which was described in detail by Gähwiler et al. [9]. This method results in a dramatic thinning of the slice over time resulting in sacrifice of three-dimensional cytoarchitecture. Another classic technique for culturing CNS tissue utilized Maximov-type chambers rather than rolling test tubes, in order to achieve "organotypic" cultures that were many cell layers thick [10]. While Roller-tube and Maximov-chamber methodologies offered distinct advantages over culturing primary cells, these ex vivo culturing systems have fallen out of favor because such preparations are technically cumbersome, provide only limited access to slices/medium, and often result in experimental variability [11].

The original membrane-interface method of brain slice culture [12] was brought into practical use by Stoppini et al. [13]. In short, this procedure involves placement of explanted rodent brain slices upon a semiporous membrane insert. The insert, itself, sits in a medium-containing well, such that the slices are suspended at the interface between medium and a humidified atmosphere (36 °C, 5 % CO_2) (*see* Fig. 1). A thin film of medium is formed above each slice through capillary action, allowing a sufficient level of hydration and nutrient absorption (without sacrifice of gas exchange) for successful maintenance of tissue survival over several weeks in culture [14]. Slices become progressively thinner over this time span, but remain "organotypic" and will not reach the minimal thickness attained via Roller-tube preparation.

Underscoring the impact it has had on modern neuroscience, the Stoppini et al. [13] membrane-interface culturing method has been utilized with steadily increasing frequency over the last two decades and has been cited in over 2,000 primary publications to date. The popularity of membrane-interface cultured slice methodology is largely due to the fact that in vivo neurobiology (i.e., neuronal differentiation, dendritic arborization, spine formation, neurotransmitter release, and receptor distribution) is remarkably well replicated in these systems and, as a result, local synaptic circuitry (mini currents, excitatory postsynaptic potentials [EPSPs], inhibitory postsynaptic potentials [IPSPs], long-term potentiation [LTP], long-term depression [LTD]) remains functionally intact [13, 15–20]. The utility of membrane-interface slice models is probably best demonstrated by pharmacological and genetic studies that yield similar experimental results in both ex vivo and in vivo settings [21–23]. Unlike the in vivo CNS, however, the extracellular environment (tissue surface, medium, and atmosphere) is readily accessible when culturing on membrane-interface inserts. This allows unmatched control over experimental conditions (physical, chemical, and electrical) and precise monitoring of medium and tissue parameters over extended time periods.

In addition to yielding mechanistic insight into fundamental aspects of normal neurobiology, such as learning/memory [16], brain development [17, 18], and neurogenesis [24], organotypic slice culture techniques have been utilized to develop highly relevant systems for the in vitro study of neurologic disease. Insults to cultured brain slices, such as oxygen-glucose deprivation (OGD), mechanical disruption, and pharmaceutical application, have become established models for studying stroke [21, 25], traumatic brain injury [26–28], and epilepsy [14, 29, 30]. Multiple ex vivo slice models of neurodegenerative disease, including Alzheimer's and Parkinson diseases, have been developed [31–35]. In addition, bacterial [36, 37], parasitic [38, 39], and viral [40–45] infections of the CNS have been modeled in slice cultures.

Not surprisingly, several of the established slice models of neurologic disease have become valuable in the drug discovery process. Pharmaceuticals can be directly applied to slices or slice medium to initially screen for efficacy, in the absence of complications related to the drug's metabolic stability or brain penetration. Given that multiple brain slices can be made from a single animal, ex vivo pharmaceutical screening is relatively high-throughput compared to whole animal studies. It is also relatively high-content when compared to cell culture studies, in that a positive "hit" is more likely to translate to animals because the system is inherently more contextual to in vivo tissue and disease biology [46].

Methods for the production and long-term culturing of rodent brain and spinal cord slices are outlined in the following sections. We also describe methods for viral infection of slice cultures and subsequent assessment of injury and apoptosis. Notably, injury can

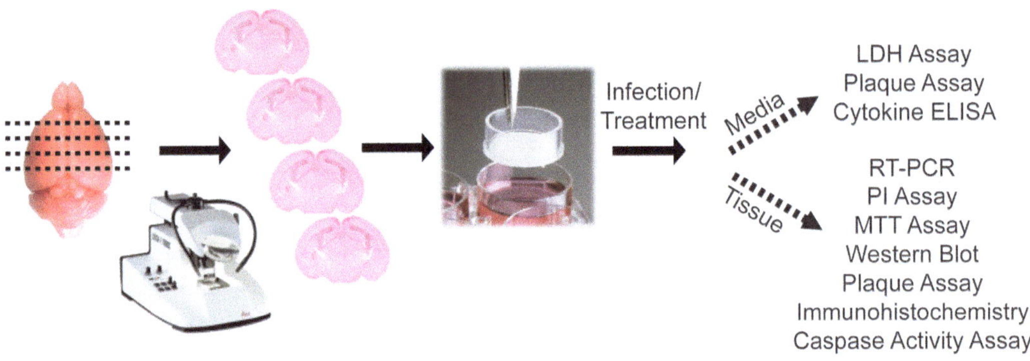

Fig. 2 Possible workflow for production of CNS tissue slices, subsequent infection/treatment, and multiple endpoints. Analysis of media is useful for the quantification of tissue injury, virus release, and cytokine release and does not require destruction of slice tissue

be assessed over time in a single sample via detection of released lactate dehydrogenase (LDH) in feeding medium. In addition, ELISA-based quantification of cytokine release into the medium is presented, as a sensitive technique that allows investigation of the innate immunological response mounted by the CNS during viral infection. In this system, CNS tissue is completely isolated from systemic, cell-mediated immunity, thus, measured cytokine responses are considered "CNS-specific". Finally, cytokine and pharmaceutical application to slice cultures are discussed. A possible work flow diagram is depicted in Fig. 2, though these methods can readily be customized according to the questions posed by individual investigators.

2 Materials

2.1 Slice Preparation (See Note 1)

2.1.1 Brain

1. 2–3 day old mice (*see* **Note 2**).
2. Razor blades.
3. Surgical tools for dissection: large scissors, small scissors, large forceps, small forceps.
4. Tools for slice manipulation: weighing spatula (*see* **Note 3**) and paintbrush.
5. Vibratome (Leica; Buffalo Grove, IL).
6. Slicing medium: MEM supplemented with 10 g/L D-glucose and 1 mM HEPES (pH 7.2). Can be stored at 4 °C for 3 weeks. Optional: equilibrate medium with 95 % O_2/5 % CO_2 immediately prior to use.
7. 10 % FBS plating medium: Neurobasal supplemented with 10 mM HEPES, 1× B-27, 400 μM L-glutamine, 600 μM GlutaMAX, 60 U/mL penicillin, 60 μg/mL streptomycin, 6 U/mL nystatin, 10 % FBS.

8. Membrane inserts (Millipore #PICMORG50, Billerica, MA).
9. 6-well cell culture plates.
10. Humidified incubator set at 5 % CO_2 and 36.5 °C.
11. Laminar flow hood (optional).
12. 95 % O_2/5 % CO_2 tank (optional).

2.1.2 Spinal Cord

1. 4–5 day old mice (*see* **Note 2**).
2. 5 mL syringe with Luer lock tip.
3. 26G needle.
4. Ca^{2+}- and Mg^{2+}-free PBS.
5. Agarose slicing medium: MEM supplemented with 10 g/L D-glucose, 1 mM HEPES (pH 7.2), and 2 % low melt agarose (Bioexpress, E-3112; Kaysville, UT).
6. Disposable plastic cryomolds.

2.2 Viral Infection

1. Purified viral stock.
2. Ca^{2+}- and Mg^{2+}-free PBS.

2.3 Culture Maintenance

1. 5 % FBS plating medium: Neurobasal supplemented with 10 mM HEPES, 1× B-27, 400 μM L-glutamine, 600 μM GlutaMAX, 60 U/mL penicillin, 60 μg/mL streptomycin, 6 U/mL nystatin, 5 % FBS.
2. Serum free plating medium: Neurobasal supplemented with 10 mM HEPES, 1× B-27, 400 μM L-glutamine, 600 μM GlutaMAX, 60 U/mL penicillin, 60 μg/mL streptomycin, 6 U/mL nystatin.
3. Humidified incubator set at 5 % CO_2 and 36.5 °C.

2.4 RT-PCR Quantification of Viral or Host Genes

1. Ca^{2+}- and Mg^{2+}-free PBS.
2. RLT buffer (Qiagen; Germantown, MD).
3. β-mercaptoethanol.
4. QIAshredder spin column (Qiagen; Germantown, MD).
5. RNeasy spin kit (Qiagen; Germantown, MD).
6. Tabletop centrifuge.
7. Primers to gene/s of interest and housekeeping gene (i.e. β-actin or GAPDH).
8. iScript™ One-Step RT-PCR Kit with SYBR® Green.
9. Bio-Rad CFX96 thermocycler with compatible hardware and software (Hercules, CA).

2.5 Cryosectioning and Immuno-fluorescence Imaging

1. Ca^{2+}- and Mg^{2+}-free PBS.
2. 4 % paraformaldehyde (PFA).
3. 15 % sucrose in PBS.
4. 30 % sucrose in PBS.
5. Tissue Freezing Medium (Triangle Biomedical Sciences; Durham, NC) (*see* **Note 4**).
6. Metal cylinder (~1 in. diameter).
7. Aluminum duct tape.
8. Cryostat.
9. ColorFrost™ Plus microscope slides (Fisher; Pittsburgh, PA).
10. Antigen unmasking solution (Vector Laboratories; Burlingame, CA).
11. Fetal bovine serum (FBS).
12. Normal goat serum (NGS).
13. 1° antibodies of interest.
14. Fluorescent 2° antibodies.
15. Hoechst stain.
16. ProLong® Gold mounting medium (Life Technologies; Grand Island, NY).
17. Epifluorescence microscope and compatible imaging software.

2.6 Propidium Iodide (PI) Quantification of Tissue Injury

1. Propidium iodide (Sigma; St. Louis, MO).
2. Dounce homogenizer.
3. 96-well plate.
4. Plate centrifuge.
5. Spectrofluorometer.

2.7 Lactate Dehydrogenase (LDH) Quantification of Tissue Injury

1. LDH-Cytotoxicity Assay Kit II (Biovision; Mountain View, CA).
2. Microplate reader.

2.8 MTT (3-(4,5-Dimethyl-thiazol-2-yl)-2,5-diphenyltetrazolium bromide) Visualization of Tissue Injury

1. MTT (Life Technologies; Grand Island, NY).
2. Digital camera.

2.9 Western Blotting

1. Ca^{2+}- and Mg^{2+}-free PBS.
2. Lysis Buffer (1 % Triton-X, 10 mM triethanolamine–HCl, 150 mM NaCl, 5 mM EDTA, 1 mM PMSF, 1× Halt protease and phosphatase inhibitor cocktail [Thermo Scientific; Rockford, IL]).
3. Tabletop centrifuge.
4. Polyacrylamide gel reagents.
5. Western blot gel apparatus.
6. Protein transfer apparatus.
7. Nitrocellulose membrane.
8. 1° antibodies of interest.
9. HRP-conjugated 2° antibodies.
10. Chemiluminescence reagent.
11. Chemiluminescence detection equipment.

2.10 Caspase Activity Assay

1. Ca^{2+}- and Mg^{2+}-free PBS.
2. Caspase fluorogenic assay kit (B&D Biosciences; San Jose, CA).
3. Spectrofluorometer.

2.11 Cytokine Analysis

1. Multi-Analyte ELISArray plate (SABiosciences, Frederick, MD).
2. Single-Analyte ELISA plate/s (SABiosciences, Frederick, MD).
3. Microplate reader.

2.12 Pharmaceutical/Cytokine Treatment

1. Ca^{2+}- and Mg^{2+}-free PBS.
2. Pharmaceutical/s or cytokine/s of interest.

3 Methods

3.1 Brain and Spinal Cord Tissue Preparation

1. Place slicing medium on ice and, if possible, bubble slicing medium with 95 % O_2/5 % CO_2.
2. Pipette 1.2 mL 10 % FBS plating medium into each well of 6-well plates. Place plate into incubator (36.5 °C, 5 % CO_2) for at least 1 h prior to explantation to allow the medium to warm and pH adjust.
3. Prepare razor blade for placement in Vibratome chuck by carefully removing the blade's back guard with the weighing spatula or flat-head screw driver. Spray and wipe blade with 70 % ethanol to remove manufacturing oil.
4. Mount razor blade into Vibratome chuck and assemble Vibratome. Set Vibratome settings (*see* **Note 5**).

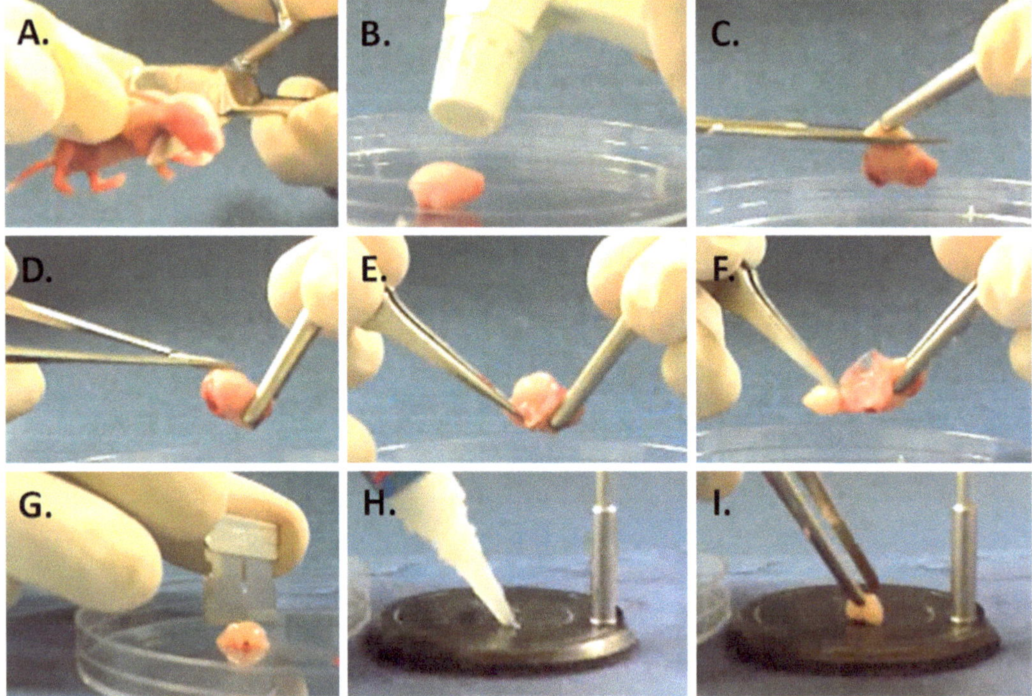

Fig. 3 Steps to harvest brains and mount to Vibratome stage. (**a**) Rapid decapitation is followed by (**b**) ethanol rinse of head. (**c**) Scalp is removed by tenting skin and making a transverse cut. (**d**) The skull is hemisected starting from foramen magnum and moving rostrally. (**e**) Each half of skull is peeled away from underlying brain and (**f**) the brain is lifted from the cranium. (**g**) Cerebrum is separated from cerebellum with razor blade. (**h**) A glue spot of approximately the same size as the freshly cut cerebral surface is placed on the Vibratome stage and (**i**) the cerebrum is placed on the glue spot, with olfactory bulbs oriented upwards

5. Clean all tools and Vibratome with 70 % ethanol. If possible, sterilize equipment with ~20 min exposure to UV light in laminar flow hood.
6. Obtain mice.
7. Harvest mouse brains or spinal cords (*see* **Note 6**).
   (a) Brains (*see* Fig. 3a–g): rapidly decapitate 2–3 day old animal with large scissors. Spray the head with 70 % ethanol. Tent the scalp with large forceps and remove with a single transverse cut made with large scissors. Stabilize the anterior skull with large forceps placed between eyes and nose. Insert small dissection scissors into foramen magnum and carefully cut the skull as far anteriorly as possible (point the scissors upward toward the skull to avoid damaging the underlying brain). Use the small forceps to carefully pry each half of the skull away from the brain. Then gently lift the brain out of the skull, severing the nerve tracts in the process. Place the brain in a Petri dish. Use a new

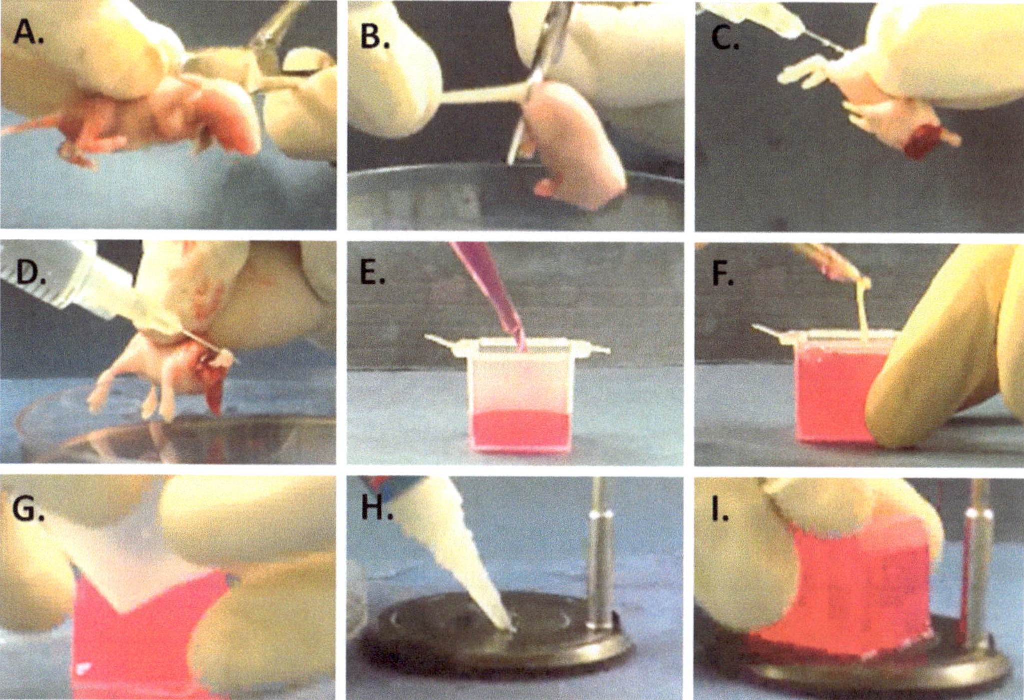

Fig. 4 Steps to harvest spinal cords, agarose embed, and mount to Vibratome stage (**a**) Rapid decapitation is followed by ethanol rinse of the body. (**b**) The tail is removed at its most proximal point. (**c**) The bevel of a 25G is inserted into the spinal column and a stream of PBS is forcefully injected through the column, impelling the spinal cord from the anterior body. (**d**) The spinal cord is gently teased away from the body. (**e**) Upon collecting several cords, pour a 2 % agarose form (**f**) and transfer each cord into the filled form. (**g**) The agarose is solidified and removed from the form. (**h**) A glue spot of approximately the same size as the freshly cut agarose block is placed on the Vibratome stage and (**i**) the widest base of the agarose block is placed on the glue spot, with the spinal cords oriented perpendicular to the Vibratome stage

razor blade to make a single coronal cut to separate cerebrum from cerebellum. This forms a flat surface on the cerebrum that can be glued to the Vibratome stage.

(b) Spinal cords (*see* Fig. 4a–g): rapidly decapitate 4–5 day old animal with large scissors. Spray the body with 70 % ethanol. Cut the tail away from the body at its most proximal point (the sacral spine should be severed in the process). Insert the 26G needle into the spinal column, such that the bevel is just buried (*see* **Note 7**). With steady force, push a PBS stream through the spinal column until the spinal cord is forced out of the spinal column. The spinal cord often remains connected at the anterior portion of the body and can be gently manipulated into a Petri dish containing slice medium or PBS. Upon collecting multiple spinal cords, pour 37 °C, 2 % low melt agarose into a cryoform. Immediately, suspend spinal

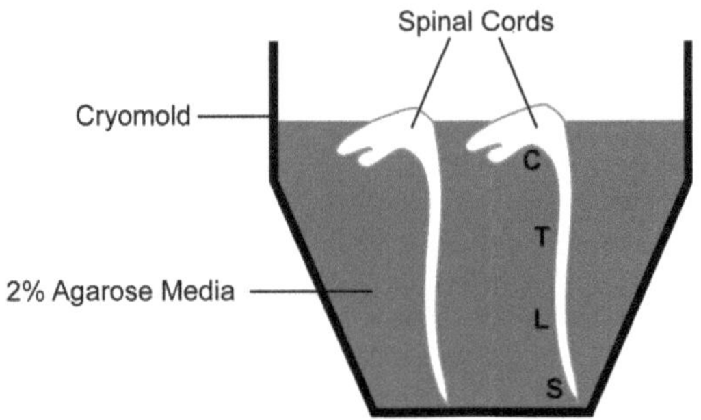

Fig. 5 Schematic of spinal cord placement in 2 % agarose. Cutaway view depicts a cryomold containing spinal cords, which are supported by the density of agarose. Cervical (C) spinal cord provides buoyancy so that thoracic (T), lumbar (L), and sacral (S) spinal cord regions are hung vertically

cords in the agarose, with cervical spinal cord oriented upwards. Manipulate the cords such that they are suspended upright in the block with the cervical cord providing buoyancy (*see* Fig. 5). Allow the block to cool/solidify (*see* **Note 8**).

8. Glue specimen to Vibratome stage (*see* **Note 9**).
   (a) Brains (*see* Fig. 3h, i): Place a small amount of super glue onto the stage and place the brain onto the glue, with frontal lobes oriented upwards (newly cut, flat side down on the glue spot). Wait for ~30 s for glue to dry, place the stage inside the Vibratome bath, then pour ice-cold slicing medium into the Vibratome bath until the point that the olfactory bulbs/frontal lobe are just submerged.
   (b) Spinal cords (*see* Fig. 4h, i): Carefully remove the agarose block from the cryomold. The block can be trimmed with a razor blade. Place a small amount of super glue onto the stage and place the agarose block onto the glue, with sacral regions oriented upward and cervical regions oriented closest to the stage. Wait for ~1 min for glue to dry, place the stage inside the Vibratome bath, then pour ice-cold slicing medium into the Vibratome bath until the point that the agarose block is just submerged.
9. Slicing (*see* **Notes 10** and **11**).
   (a) Brains: Initial coronal slices will be of olfactory bulbs, followed by slices of frontal lobes. We generally discard these slices. When the bi-lobar shape of the frontal lobes is no longer apparent, sections will contain both hippocampus and thalamus. We generally save these slices for culture. Four, 400 μM slices containing hippocampus/thalamus can be obtained from each 2–3 day old mouse brain.

By gentle pickup with a paintbrush and bent weighing spatula, collect each nascent slice into 6-well plates containing ice-cold slicing medium. If further dissection of tissue slices is desired *see* **Note 12**.

(b) Spinal cords: Initial transverse slices will be of sacral and lumbar sections, followed by thoracic and cervical sections. We generally discard sacral sections (too small to work with) and cervical slices (not true transverse slices). Approximately 20, 400 μM slices can be obtained from each 4–5 day old mouse. By gentle pickup with a paintbrush and bent weighing spatula, collect each nascent agarose slice into a Petri dish containing ice-cold, oxygenated slicing medium. Individual spinal cord slices can easily be dislodged from agarose by holding the agarose slice in a stationary position and gently "blowing" slicing medium toward the slices with a disposable pipette.

3.2 Plating Brain and Spinal Cord Slices

1. Place a Millipore insert into each of the 6-well plate wells with sterile forceps.
2. Carefully transfer slices onto the Millipore insert.
   (a) Brain slices (*see* **Note 13**): Gently coax the slice onto the bent weighing spatula with the paintbrush. Once secure, place the spatula above the insert at a steep angle. Pick up a large droplet of slicing medium by briefly submerging the paintbrush, then place the droplet at the higher end of the spatula. Ideally the droplet will roll down the incline of the spatula and wash the slice onto the membrane. It is important that the slices lay perfectly flat on the membrane.
   (b) Spinal cord slices: Using a disposable pipette, pick up slices (and minimal slicing medium) and transfer to the membrane.
3. Remove any slicing medium that is transferred along with the slices by Pasteur pipette. The slices should not be submerged in large volumes of liquid because slices need exposure to incubator atmosphere for optimal survival.
4. Place plates in humidified incubator set at 5 % CO_2 and 36.5 °C.

3.3 Viral Infection

1. Immediately after plating or following slice recovery (*see* Fig. 6) add purified virus (diluted in PBS) dropwise to the top (air interface) of each slice. A volume of 15–20 μL covers a single brain slice and a volume of 5–10 μL covers a single spinal cord slice. Minimized inoculation volumes will ensure that slices are not submerged in liquid for long periods of time. Small volumes will be absorbed into the slice tissue and enter the underlying medium in <5 min.

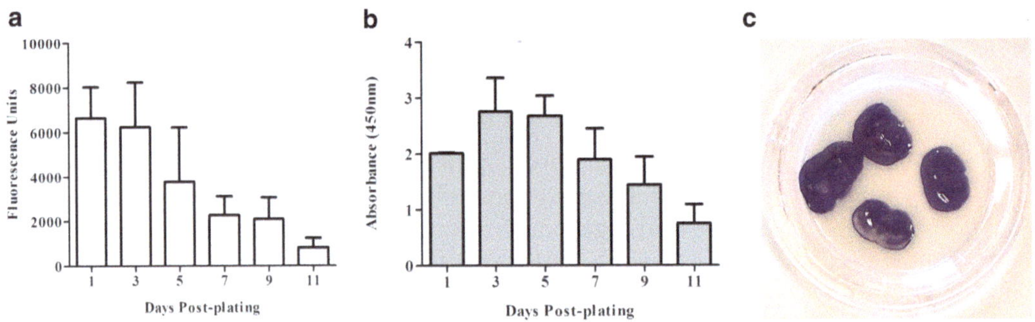

Fig. 6 Prior to experimentation, CNS slices may need time to recover from axonal/cellular injury induced by the slicing procedure. (**a**) Brain slices were harvested into lysis buffer at indicated times post-plating ($N=3$–4). To quantify apoptosis, Caspase 3 activity was determined by fluorogenic activity assay. Apoptosis occurs during the first week in culture, but is negligible at later time points. (**b**) Media was taken at specified time points ($N=3$–12). To quantify tissue injury, LDH was detected by colorimetric assay. Tissue injury is evident in the week following slicing, but then progressively decreases over time. (**c**) At 11 days post-plating, BSC media was replaced with fresh media containing MTT and a representative photograph was taken 45 min later. Live tissue cleaves the substrate to form *purple* formazan crystals through mitochondrial activity. This demonstrates the long-term viability of brain slice tissue, despite initial apoptosis and tissue injury caused by slicing

2. Mock inoculations should be performed in a similar manner with PBS alone.
3. Time points need to be determined by each individual investigator. If in vivo time points are known, these should be used as starting points. It is recommended that all experimentation is completed prior to the 21st day post-plating.

3.4 Culture Maintenance (See Note 14)

1. At ~12 h post-plating, pipette 1.2 mL 5 % FBS plating medium into each well of new 6-well plates. Place plates into incubator (36.5 °C, 5 % CO_2) for at least 1 h prior to slice transfer to allow the medium to warm and pH adjust.
2. With sterile forceps, transfer each membrane insert into wells containing fresh medium.
3. Replace medium every 2 days thereafter with serum-free medium.

3.5 RT-PCR Quantification of Viral or Host Genes

3.5.1 Harvest Slices Into RLT-βME Buffer

1. Add 10 μL β-mercaptoethanol to each 1 mL of RLT buffer.
2. Pipette 600 μL of RLT-βME buffer into labeled 1.5 mL microcentrifuge tubes.
3. Rinse slices two times by briefly submerging the membrane insert into the wells of a 6-well plate filled with PBS.
4. Lift slices away from the membrane by gently sweeping under slices with the weighing spatula (*see* **Note 15**). Put up to four brain slices or 20 spinal cord slices directly into 600 μL of RLT-βME buffer. Triturate the tissue through a pipette tip until it is completely dissociated into the buffer.
5. Freeze the lysate at −80 °C or proceed immediately to the next step.

3.5.2 Purify Slice RNA

1. Homogenize the lysate by pipetting it into a QIAshredder and spinning at 12,000 × *g* for 2 min.
2. Add 600 μL of 70 % ethanol (in DEPC water) to each homogenized lysate and mix by gentle trituration through a pipette tip.
3. Pipette half of the homogenized lysate/ethanol (600 μL) into a labeled Qiagen RNeasy MINI spin column.
4. Spin at 12,000 × *g* for 1 min, discard flow through.
5. Repeat **steps 3** and **4** with the remainder of the sample.
6. Add 700 μL of RW1 buffer to the column, spin at 12,000 × *g* for 15 s, discard flow though.
7. Add 500 μL of RPE buffer, spin at 12,000 × *g* for 15 s, discard flow though.
8. Add another 500 μL RPE, spin at 12,000 × *g* for 2 min, discard flow though.
9. Spin again at 12,000 × *g* for 1 min to dry column.
10. Carefully place spin column to a new 1.5 mL microcentrifuge tubes. Add 50 μL of water to each column.
11. Wait 1 min and spin again at 12,000 × *g* for 1 min.
12. Store purified RNA at −80 °C.

3.6 RT-PCR Quantification

1. Design and synthesize primers to gene/s of interest and a housekeeping gene.
2. Mix purified RNA template, primers, SYBR® Green RT-PCR master mix, and iScript™ reverse transcriptase into a total volume of 20 μL.
3. Perform 40 cycles of PCR amplification on thermocycler as follows: cDNA synthesis at 50 °C for 10 min, reverse transcriptase inactivation at 95 °C for 5 min, denaturation at 95 °C for 10 s, and annealing/extension at 60 °C for 30 s.
4. Check melt curve to confirm the absence of nonspecific products and primer dimers.
5. Convert raw *C*(t) values into relative expression values with compatible analysis software.

3.7 Cryosectioning and Immunofluorescence Imaging (See Note 16)

3.7.1 Fix and Cryoprotect Tissue

1. Rinse slices two times by briefly submerging the membrane insert into the wells of a 6-well plate filled with PBS.
2. Transfer membranes into a 6-well plate filled with 4 % PFA and allow ~12 h fixation at 4 °C (make sure slices are completely submerged in fixative).
3. Rinse slices two times by briefly submerging the membrane insert into the wells of a 6-well plate filled with PBS.
4. Lift slices away from the membrane by gently sweeping under slices with the weighing spatula (*see* **Note 15**). Put loose slices directly into 15 % sucrose for cryoprotection.

5. Store slices for ~1 day in 15 % Sucrose at 4 °C to cryoprotect tissue.
6. Store slices for ~1 day in 30 % Sucrose at 4 °C to cryoprotect tissue.

3.7.2 Form a "Face-Off" TFM Block (Used to form a Surface That Is Parallel to Cryostat Blade)

1. Create a small well by running aluminum tape around the end of a metal cylinder, such that the tape edge is higher than the edge of the cylinder.
2. Fill well with TFM and place metal cylinder in dry ice until TFM solidifies (this forms a "face-off" block that does not contain slices).
3. Take the cylinder out of dry ice, remove the tape, remove the "face-off" block from the metal cylinder, and store the "face-off" block in aluminum foil at −80 °C.

3.7.3 Cryoembed Slices

1. Place slices in room temperature TFM for at least several minutes.
2. Carefully lift slices out of room temperature TFM with a paint-brush and place them in a metal well identical to that created in Subheading **3.7.2**. The slice should be laid flat against the surface of the metal cylinder, without the creation of bubbles.
3. Gently apply a thin layer of TFM over the tissue slices and place the metal cylinder in dry ice until TFM solidifies (this forms a thin block that contains slices at the extreme bottom).
4. Take the cylinder out of dry ice, remove the tape, remove the "slice block" from the metal cylinder, and store the "slice block" in aluminum foil at −80 °C.
5. Repeat until all tissue slices have been cryoembedded in "slice blocks."

3.7.4 Cryosectioning

1. Attach the "face-off block" to the cryostat chuck and make sections until the entire surface of the block is removed with each pass of the blade (i.e., the entire block face is perfectly parallel to the blade path).
2. Mark orientation of the "face-off block" with a marker.
3. Remove the "face-off block" from the cryostat chuck.
4. Coat the "face-off block" with a thin layer of TFM and immediately place a "slice block" (with slices oriented downward, so that the slices are sandwiched inside the two blocks).
5. Attach the "sandwich block" to the cryostat chuck.
6. Make sections until the slices are clearly visible (*see* **Note 4**).
7. Ensure that the blade is sharp and recent sections are of high quality.

8. Make serial 10–20 μm sections through the tissue and mount on ColorFrost™ Plus slides (*see* **Note 17**).
9. Store slides at room temperature until immunohistochemical processing.

3.7.5 Immunohistochemistry

1. Desiccate slides at 50 °C for 15 min.
2. Rehydrate slides in PBS over 30 min.
3. Perform antigen retrieval with antigen unmasking solution according to manufacturer's instructions.
4. Permeabilize tissue and block antigen by submerging tissue with 5 % fetal bovine serum and 5 % normal goat serum in 0.3 % Triton/PBS for 2 h at room temperature.
5. Incubate with primary antibodies overnight at 4 °C.
6. Incubate with secondary antibodies for 2 h at room temperature.
7. Stain nuclei with Hoechst stain according to manufacturer's instructions.
8. Mount sections with ProLong® Gold mounting medium.
9. Image slides on an epifluorescence microscope.

3.8 Propidium Iodide (PI) Quantification of Tissue Injury

1. Replace serum-free medium, with serum-free medium containing 3 μM propidium iodide.
2. Return cultures to incubator for a 1 h incubation.
3. Rinse slices two times by briefly submerging the membrane insert into the wells of a 6-well plate filled with PBS.
4. Homogenize four brain slices or spinal cord slices into 500 μL PBS with a Dounce homogenizer.
5. In triplicate, pipette 150 μL of homogenate into 96-well plate.
6. Spin down cells at 1,000 × *g* for 5 min.
7. Desiccate cells by incubating in a dry incubator set at 37 °C for 24 h.
8. Quantify tissue injury with a fluorescence plate reader (excitation wavelength of ~493 nm and emission wavelength of ~630 nm).

3.9 Lactate Dehydrogenase (LDH) Quantification of Tissue Injury

1. Take samples (≥40 μL) of feeding medium at sequential time points.
2. According to manufacturer's instructions, detect LDH in triplicate 10 μL samples by colorimetric assay and quantification on a 96-well microplate reader.

3.10 MTT Visualization of Tissue Injury

1. Replace serum-free medium, with serum-free medium containing 0.5 mg MTT/mL medium.
2. In 30 min to 1 h, take photographs to characterize regionality of tissue injury (blue/purple indicates live tissue, whereas white/brown indicates otherwise).

3.11 Western Blotting

1. Pipette lysis buffer (50 μL/brain slice or 10 μL/spinal cord slice) into labeled 1.5 mL microcentrifuge tubes.
2. Rinse slices two times by briefly submerging the membrane insert into the wells of a 6-well plate filled with PBS.
3. Lift slices away from the membrane by gently sweeping under slices with the weighing spatula (*see* **Note 15**). Put slices directly into lysis buffer. Triturate the tissue through a pipette tip until it is completely dissociated into the buffer.
4. Freeze the lysate at −80 °C or proceed immediately to the next step.
5. If frozen, place homogenates on ice until thawed. Then allow cell lysis to occur for 5–10 min, while vortexing frequently.
6. Clear lystates, by spinning at 16,000 × *g* for 10 min at 4 °C.
7. Transfer supernatant to fresh tube containing equal volume of 2× Laemmli buffer.
8. Boil for 5 min.
9. Load and run samples on polyacrylamide gel (*see* **Note 18**).
10. Transfer to nitrocellulose membrane.
11. Block membrane with 5 % milk.
12. Immunoblot with primary antibody overnight at 4 °C.
13. Wash membrane and incubate with appropriate HRP-conjugated secondary antibody.
14. Obtain film or digital images following addition of chemiluminescence reagent.

3.12 Caspase Activity Assay

1. Rinse slices two times by briefly submerging the membrane insert into the wells of a 6-well plate filled with PBS.
2. Lift slices away from the membrane by gently sweeping under slices with the weighing spatula (*see* **Note 15**). Put slices directly into lysis buffer provided by manufacturer. Triturate the tissue through a pipette tip until it is completely dissociated into the buffer.
3. Freeze the lysate at −80 °C or proceed immediately to the next step.
4. Quantify caspase activity according to manufacturer's instructions.

3.13 Cytokine Analysis

3.13.1 Cytokine Screen on Multi-Analyte ELISArray Plates

1. Collect samples of medium at sequential time points.
2. Pool medium from experimentally similar wells together for cytokine screening (it is suggested that some medium is left unpooled, such that medium from individual wells can later be used for cytokine quantification on single-analyte ELISA plates).
3. Screen pooled medium on multi-analyte ELISArray plates according to manufacturer's instructions.

3.13.2 Cytokine Quantification Single-Analyte ELISA Plates

1. Obtain single-analyte ELISA plates for each cytokine that was positive on the ELISArray.
2. With the un-pooled medium collected in Subheading **3.13.1**, quantify cytokine of interest in each medium sample through serial dilutions fit to a standard curve according to manufacturer's directions.

3.14 Pharmaceutical/Cytokine Treatment

1. Obtain in vitro dosage information, if available.
2. Dilute pharmaceutical/cytokine so that it may be applied directly to the top (air interface) of each slice. The initial concentration must be non-toxic and should contain ≤10 % DMSO. A volume of 15–20 μL covers a single brain slice and a volume of 5–10 μL covers a single spinal cord slice. Small volumes will be absorbed into the slice tissue and enter the underlying medium in ~5 min. Ideally, the final concentration should be above the effective concentration as established in vitro.
3. Reapply pharmaceutical/cytokine following each medium change.

4 Notes

1. Slices are vulnerable to fungal infection. During preparation, we recommend wearing a mask and sterilizing gloves frequently with 70 % ethanol. Nystatin is added to medium to alleviate problems with fungal infection.
2. Slices can be prepared from both mice and rats using described methodology. While we recommend using young postnatal animals, due to the resilience of tissue at this age, other investigators have successfully cultured CNS tissue derived from animals that are several weeks old at the time of sacrifice. In our experience, P2–3 seems ideal for the brain slices model and P4–5 seems ideal for the spinal cord model.
3. Bend the end of the spatula to a 45° angle for easier manipulation of slices.
4. Usage of colored TFM (red, green, blue) helps tremendously with the visualization of slices during cryosectioning.

5. Our laboratory uses the following settings on a Leica VT1000S Vibratome: advance speed = 7–10 and amplitude = 7–10. Exact settings need to be determined by individual investigators. Our brain and spinal cord slices are made at 400 μm thickness. Slices can be made thinner (i.e., 250 μm), though thin slices will have a larger percentage of the total slice damaged by the slicing procedure. It is not advisable to make slices thicker than 400 μm, because the top of thick slices will not have adequate contact with feeding medium.
6. It is important that this step be performed quickly. With practice, it should not take any longer than 5 min and up to six brains or spinal cords may be harvested and sliced at once.
7. Aim for the small blood spot that immediately forms when the sacral spinal cord is severed. Gentle side-to-side movement of the needle can be used to confirm that it has indeed entered the spinal column. Furthermore, correct needle placement is confirmed if limb flexion is induced by mechanical stimulation of the spinal cord when PBS is forced through the spinal column. Incorrect needle placement, on the other hand, will cause the body cavity to fill with PBS.
8. Cooling of the agar block can be hastened by briefly placing the form on ice or at 4 °C.
9. Size of glue spot should be approximately the same size as the "footprint" of the brain or spinal cord-containing agarose block. Controlling the glue volume is a critical part of the slicing procedure: if too little glue is applied the specimen will become dislodged from the Vibratome stage, whereas too much glue will coat the outside of the specimen, harden, and induce damage to the tissue and razor blade.
10. Replace the razor blade often (i.e., every time the stage is loaded with new specimens). Small imperfections in the razor blade can result in unnecessary damage to tissue slices, whereas, usage of a sharp blade will minimize tissue tearing.
11. Use a paintbrush to gently hold the nascent slice against the blade; this keeps it from rolling up and allows for a cleaner cut.
12. If isolated hippocampal cultures are desired, individual hippocampi may be dissected out of nascent slices with a dissecting scope prior to plating. If many experimental conditions will be applied (e.g., higher throughput drug screening), each slice may be split into right and left hemispheres with a scalpel prior to plating.
13. The plating procedure for brain slices is technically challenging. Accordingly, mastery of this step will take significant patience.
14. We prefer to culture in a serum-free system to minimize variables. However, providing FBS to nascent slices helps with

tissue recovery. So, we tend to initially plate cultures with 10 % FBS, and wean to 5 % FBS at ~12 h post-plating, then switch to serum-free medium at 3 days post-plating.

15. Separating slices from culture membranes will become more difficult with increased time in culture due to glia incorporating into the membrane.
16. As an alternative to labor-intensive cryosectioning and epifluorescence imaging, confocal imaging of the surface of the whole slice may be performed. Confocal microscopy also makes live imaging possible.
17. Tissue will not bind well to uncoated slides and will wash off during immunofluorescence processing. We recommend usage of coated slides.
18. In our experience, most proteins are detectable (yet not overloaded) when the equivalent of one brain slice or five spinal cord slices are loaded per well.

Acknowledgements

The authors are supported by RO1NS076512 (KLT), R21AI01064 (KLT), VA Merit Grant BX000963 (KLT), T32GM008497 (KRD), and F30NS071630 (KRD).

References

1. Hauser S (2006) Harrison's neurology in clinical medicine. McGraw-Hill Companies, New York
2. Hilgetag CC, Barbas H (2009) Are there ten times more glia than neurons in the brain? Brain Struct Funct 213:365–366
3. Kandel E, Schwartz J, Jessel T (2000) Principles of neural science, 4th edn. McGraw-Hill Companies, New York
4. Harrison RG (1907) Observations on the living developing nerve fiber. Proc Soc Exp Biol Med 4:140–143
5. Geller HM (1976) Effects of some putative neurotransmitters on unit activity of tuberal hypothalamic neurons in vitro. Brain Res 108:423–430
6. Costero I, Pomerat C (1951) Cultivation of neurons from the adult human cerebral and cerebellar cortes. Am J Anat 89:405–467
7. Storts RW, Koestner A (1968) General cultural characteristics of canine cerebellar explants. Am J Vet Res 29:2351–2364
8. Hogue M (1947) Human fetal brain cells in tissue cultures; their identification and motility. J Exp Zool 106:85–107
9. Gahwiler BH (1981) Organotypic monolayer cultures of nervous tissue. J Neurosci Methods 4:329–342
10. Crain SM (1966) Development of "organotypic" bioelectric activities in central nervous tissues during maturation in culture. Int Rev Neurobiol 9:1–43
11. Cho S, Wood A, Bowlby MR (2007) Brain slices as models for neurodegenerative disease and screening platforms to identify novel therapeutics. Curr Neuropharmacol 5:19–33
12. Yamamoto N, Kurotani T, Toyama K (1989) Neural connections between the lateral geniculate nucleus and visual cortex in vitro. Science 245: 192–194
13. Stoppini L, Buchs PA, Muller D (1991) A simple method for organotypic cultures of nervous tissue. J Neurosci Methods 37:173–182
14. Holopainen IE (2005) Organotypic hippocampal slice cultures: a model system to study basic cellular and molecular mechanisms of neuronal cell death, neuroprotection, and synaptic plasticity. Neurochem Res 30:1521–1528
15. Zimmer J, Gahwiler BH (1984) Cellular and connective organization of slice cultures of the

rat hippocampus and fascia dentata. J Comp Neurol 228:432–446
16. Collin C, Miyaguchi K, Segal M (1997) Dendritic spine density and LTP induction in cultured hippocampal slices. J Neurophysiol 77:1614–1623
17. De Simoni A, Yu LM (2006) Preparation of organotypic hippocampal slice cultures: interface method. Nat Protoc 1:1439–1445
18. Gahwiler BH (1984) Development of the hippocampus in vitro: cell types, synapses and receptors. Neuroscience 11:751–760
19. Finley M, Fairman D, Liu D et al (2004) Functional validation of adult hippocampal organotypic cultures as an in vitro model of brain injury. Brain Res 1001:125–132
20. De Simoni A, Griesinger CB, Edwards FA (2003) Development of rat CA1 neurones in acute versus organotypic slices: role of experience in synaptic morphology and activity. J Physiol 550:135–147
21. Cho S, Liu D, Fairman D et al (2004) Spatiotemporal evidence of apoptosis-mediated ischemic injury in organotypic hippocampal slice cultures. Neurochem Int 45: 117–127
22. Pringle AK, Sundstrom LE, Wilde GJ et al (1996) Brain-derived neurotrophic factor, but not neurotrophin-3, prevents ischaemia-induced neuronal cell death in organotypic rat hippocampal slice cultures. Neurosci Lett 211:203–206
23. Ray AM, Owen DE, Evans ML et al (2000) Caspase inhibitors are functionally neuroprotective against oxygen glucose deprivation induced CA1 death in rat organotypic hippocampal slices. Brain Res 867:62–69
24. Raineteau O, Rietschin L, Gradwohl G et al (2004) Neurogenesis in hippocampal slice cultures. Mol Cell Neurosci 26:241–250
25. Noraberg J, Poulsen FR, Blaabjerg M et al (2005) Organotypic hippocampal slice cultures for studies of brain damage, neuroprotection and neurorepair. Curr Drug Targets CNS Neurol Disord 4:435–452
26. Morrison B III, Eberwine JH, Meaney DF et al (2000) Traumatic injury induces differential expression of cell death genes in organotypic brain slice cultures determined by complementary DNA array hybridization. Neuroscience 96:131–139
27. Adembri C, Bechi A, Meli E et al (2004) Erythropoietin attenuates post-traumatic injury in organotypic hippocampal slices. J Neurotrauma 21:1103–1112
28. Adamchik Y, Frantseva MV, Weisspapir M et al (2000) Methods to induce primary and secondary traumatic damage in organotypic hippocampal slice cultures. Brain Res Brain Res Protoc 5:153–158
29. Routbort MJ, Bausch SB, McNamara JO (1999) Seizures, cell death, and mossy fiber sprouting in kainic acid-treated organotypic hippocampal cultures. Neuroscience 94:755–765
30. Thompson SM (1993) Consequence of epileptic activity in vitro. Brain Pathol 3:413–419
31. Selkoe DJ (2008) Soluble oligomers of the amyloid beta-protein impair synaptic plasticity and behavior. Behav Brain Res 192:106–113
32. Lesne S, Koh MT, Kotilinek L et al (2006) A specific amyloid-beta protein assembly in the brain impairs memory. Nature 440:352–357
33. Wang Q, Walsh DM, Rowan MJ et al (2004) Block of long-term potentiation by naturally secreted and synthetic amyloid beta-peptide in hippocampal slices is mediated via activation of the kinases c-Jun N-terminal kinase, cyclin-dependent kinase 5, and p38 mitogen-activated protein kinase as well as metabotropic glutamate receptor type 5. J Neurosci 24:3370–3378
34. Shimizu K, Matsubara K, Ohtaki K et al (2003) Paraquat leads to dopaminergic neural vulnerability in organotypic midbrain culture. Neurosci Res 46:523–532
35. Madsen JT, Jansen P, Hesslinger C et al (2003) Tetrahydrobiopterin precursor sepiapterin provides protection against neurotoxicity of 1-methyl-4-phenylpyridinium in nigral slice cultures. J Neurochem 85:214–223
36. Gianinazzi C, Grandgirard D, Simon F et al (2004) Apoptosis of hippocampal neurons in organotypic slice culture models: direct effect of bacteria revisited. J Neuropathol Exp Neurol 63:610–617
37. Stringaris AK, Geisenhainer J, Bergmann F et al (2002) Neurotoxicity of pneumolysin, a major pneumococcal virulence factor, involves calcium influx and depends on activation of p38 mitogen-activated protein kinase. Neurobiol Dis 11:355–368
38. Scheidegger A, Vonlaufen N, Naguleswaran A et al (2005) Differential effects of interferon-gamma and tumor necrosis factor-alpha on Toxoplasma gondii proliferation in organotypic rat brain slice cultures. J Parasitol 91:307–315
39. Gianinazzi C, Schild M, Muller N et al (2005) Organotypic slice cultures from rat brain tissue: a new approach for Naegleria fowleri CNS infection in vitro. Parasitology 131:797–804
40. Dionne KR, Leser JS, Lorenzen KA et al (2011) A brain slice culture model of viral encephalitis reveals an innate CNS cytokine response profile and the therapeutic potential of caspase inhibition. Exp Neurol 228: 222–231
41. Braun E, Zimmerman T, Ben Hur T et al (2006) Neurotropism of herpes simplex virus

type 1 in brain organ cultures. J Gen Virol 87:2827–2837

42. Mayer D, Fischer H, Schneider U et al (2005) Borna disease virus replication in organotypic hippocampal slice cultures from rats results in selective damage of dentate granule cells. J Virol 79:11716–11723
43. Tsutsui Y, Kawasaki H, Kosugi I (2002) Reactivation of latent cytomegalovirus infection in mouse brain cells detected after transfer to brain slice cultures. J Virol 76:7247–7254
44. Chen SF, Huang CC, Wu HM et al (2004) Seizure, neuron loss, and mossy fiber sprouting in herpes simplex virus type 1-infected organotypic hippocampal cultures. Epilepsia 45: 322–332
45. Friedl G, Hofer M, Auber B et al (2004) Borna disease virus multiplication in mouse organotypic slice cultures is site-specifically inhibited by gamma interferon but not by interleukin-12. J Virol 78:1212–1218
46. Sundstrom L, Morrison B III, Bradley M et al (2005) Organotypic cultures as tools for functional screening in the CNS. Drug Discov Today 10:993–1000

Chapter 10

Neurospheres and Glial Cell Cultures: Immunocytochemistry for Cell Phenotyping

Amanda Parker Struckhoff and Luis Del Valle

Abstract

Cell cultures constitute an important tool for research as a way to reproduce pathological processes in a controlled system. However, the culture of brain-derived cells in monolayer presents significant challenges that obscure the fidelity of in vitro results. After a few number of passages, glial and neuronal cells begin to lose their morphological characteristics, and most importantly, their specific cellular markers and phenotype. In recent years, the discovery of Neural Progenitor Cells and the methodology to culture them in suspension maintaining their potentiality while still retaining the ability to differentiate into astrocytes, oligodendrocytes, and neurons have made significant contributions to the fields of neuroscience and neuropathology.

In the brain, progenitor cells are located in the Germinal Matrix, in the subventricular zone and play an essential role in the homeostasis of the brain by providing the source to replace differentiated cells that have been lost or damaged by different pathological processes, such as injury, genetic conditions or disease. The discovery of these Neural Stem Cells in an organ traditionally thought to have limited or no regenerative capacity has open the door to the development of novel treatments, which include cell replacement therapy. Here we describe the culture and differentiation of neural progenitor cells from Neurospheres, and the phenotyping of the resulting cells using immunocytochemistry. The immunocytological methods outlined are not restricted to the analysis of Neurosphere-derived cultures but are also applicable for cell typing of primary glial or cell line-derived samples.

Key words Neurospheres, Neural Progenitor Cells, Immunocytochemistry, Tumor Stem Cells, Differentiation

1 Introduction

The discovery of embryonic and adult stem cells has led to great interest and excitement about the therapeutic potential of these cells across medical disciplines. The two hallmark properties of stem cells are their ability to undergo self-renewal and their potentiality for differentiation into the multiple cell lineages specific for the organ from which they derive. Because of these properties, stem cells are essential for an organism to maintain tissue homeostasis and for tissue repair following injury. The features of self-renewal and pluripotentiality also make adult stem cells invaluable therapeutic tools for

Shohreh Amini and Martyn K. White (eds.), *Neuronal Cell Culture: Methods and Protocols*, Methods in Molecular Biology, vol. 1078, DOI 10.1007/978-1-62703-640-5_10, © Springer Science+Business Media New York 2013

reversing the symptoms/pathology of neurodegenerative disease, promoting tissue repair after ischemia or stroke, or as a model system to understand how to inhibit the proliferative capacity of recently identified tumor-associated stem cells.

These cells may be isolated and cultured in vitro so that they can be manipulated and studied by researchers. Neural stem cells are present in two specialized niches in the adult mammalian brain, the dentate gyrus of the hippocampus [1] and the Germinal Matrix, located in subventricular zone of the lateral ventricles [2–4]. Once removed, they can be propagated as adherent cells in traditional two-dimensional cultures or embedded in extracellular matrices, or as floating spheres called Neurospheres [5]. Culturing Neural Stem Cells in the form of Neurospheres helps to select for and preserve stem cell-like properties of isolated cells. Neural Stem Cells grown as Neurospheres can be maintained in this primitive state for many passages, creating a readily available, stable pool of Stem Cells for research. The cells can be used to ask defined questions about factors that alter stem cell properties, such as their capacity for self-renewal, or how experimental or genetic manipulations affect the neuronal or glial differentiation program.

In order for these types of experiments to be successful, researchers must be able to definitively identify the genetic and phenotypic changes resulting from the experimental manipulations. Immunocytochemistry is an invaluable tool for cell phenotyping. Neural Stem Cells and their progeny can be identified by the use of select protein biomarkers. These biomarkers are selectively expressed by stem, neuronal, or glial cells, as well as during specific stages in the differentiation process.

2 Materials

2.1 Tissue Culture Equipment

1. Tissue culture flasks and dishes.
2. Sterile polypropylene conical tubes.
3. 40 μm cell strainer.
4. Incubator, light inverted microscope.

2.2 Culture of Neurospheres

1. 0.05 % trypsin/EDTA.
2. Trypsin neutralizing solution.
3. Phosphate buffered saline.

2.3 Medium Components

1. *Neurobasal medium.*
2. *DMEM.*
3. *EGF*—Working concentration will be 20 ng/ml. Stock concentration—10 μg/ml, store as aliquots in −20 °C.
4. *bFGF*—Working concentration will be 20 ng/ml. Stock concentration—10 μg/ml, store as aliquots in −20 °C.

5. *GlutaMAX*—use 1 ml/100 ml of media (1:100 dilution).
6. *StemPro Neural Supplement*—use 2 ml/100 ml (1:50 dilution).
7. *Penicillin/streptomycin*—Working concentration—100 U/ml (1:100 dilution from 10,000 U/ml stock solution).
8. *B27supplementserum-freesupplement*—Workingconcentration—2 % (1:50 dilution).
9. *Fetal bovine serum* (*FBS*).
10. *Cilliary neurotrophic factor* (*CNTF*)—Working concentration 100 ng/ml, Stock concentration—50 μg/ml, store aliquots in -20 °C.
11. *Triiodothyronine* (*T3*)—Working concentration—30 ng/ml.
12. *Dibutyryl cAMP*—Working concentration—0.5 mM.
13. *Liberase/HBSS solution*—Liberase™ (Roche) at 100 μg/ml and DNase at 100 μg/ml, in HBSS (Hank's Balanced Salt Solution).
14. *Poly-HEMA*.

2.4 Neurosphere Growth Medium

Neurobasal medium supplemented with 20 ng/ml EGF, 20 ng/ml bFGF, 2 mM GlutaMAX, pen/strep, and 2 % StemPro Neural Supplement.

2.5 Neurosphere Differentiation Medium

Neurobasal medium supplemented with 10 % Fetal Calf Serum, 2 mM GlutaMAX, pen/strep.

2.6 Neuronal Differentiation Medium

Neurobasal medium supplemented with 2 % B27, 2 mM GlutaMAX, and pen/strep.

2.7 Oligodendrocyte Differentiation Medium

Neurobasal medium supplemented with 2 % B27, 30 ng/ml T3, 2 mM GlutaMAX, and pen/strep.

2.8 Astrocyte Differentiation Medium

Neurobasal medium supplemented with 10 % FCS supplement, 100 ng/ml CNTF, 2 mM GlutaMAX, pen/strep.

2.9 Immunocytochemistry Reagents

1. Glass coverslips and Chamber slides (2-chamber or 8-chamber).
2. Reduced Growth Factor Matrigel.
3. Paraformaldehyde (PFA) (Electron microscopy grade).
4. Methanol.

Table 1
A list of commonly used markers for immunophenotyping and the cell types which express them

Marker	Cell Type
BLBP	Neuroprogenitor
Nestin	Neuroprogenitor, BTSC
SOX2	Neuroprogenitor, BTSC
CD133	Neuroprogenitor, BTSC
BMI1	BTSC
MUSASHI	BTSC
β-tubulin, Class III	Neuroprogenitor, Neuronal
MAP-2	Neuronal
GAP-43	Neuronal
CaM kinase	Neuronal
Neurofilaments	Neuronal
GFAP	Astrocyte
S100β	Astrocyte
GLAST	Astrocyte
Glutamine Synthetase	Astrocyte
NG2	OPC
O4	OPC, Differentiated Oligodendrocyte
CNPase	OPC, Differentiated Oligodendrocyte
GalC	OPC, Differentiated Oligodendrocyte
MBP	Differentiated Oligodendrocyte
MAG	Differentiated Oligodendrocyte
MOG	Differentiated Oligodendrocyte
OX-42/CD-11b	Microglial

BTSC Brain Tumor Stem Cell, *OPC* Oligodendrocyte Progenitor Cell, *BLBP* Brain lipid-binding protein, *SOX2* SRY (sex determining region Y)-box 2 protein, *BMI1* B lymphoma Mo-MLV insertion region 1 homolog, *MAP2* Microtubule-associated protein 2, *GAP43* growth associated protein 43, *CaM kinase* Ca^{2+}/calmodulin-dependent protein kinase, *GFAP* Glial fibrillary acidic protein, *GLAST* GLutamate ASpartate Transporter, *CNPase* 2′,3′-Cyclic-nucleotide 3′-phosphodiesterase, *GalC* Galactosylceramidase, *MBP* Myelin basic protein, *MAG* Myelin-associated glycoprotein, *MOG* Myelin oligodendrocyte glycoprotein

5. Acetone.
6. Triton X-100.
7. Primary Antibodies (*see* Table 1).

8. Secondary Antibodies—all at 1:500 dilution.
   (a) Rabbit anti-mouse (Alexa Fluor488).
   (b) Goat anti-rabbit (Alexa Fluor488).
   (c) Rabbit anti-mouse (Alexa Fluor568).
   (d) Goat anti-rabbit (Alexa Fluor568).

3 Methods

3.1 Isolation of Neural Stem Cells

1. Adult-derived Neurospheres are derived from tissue isolated from the SVZ of 6–8 week old mice [6].
2. Mince tissue using the blade of a scalpel for approximately 1 min, until only small pieces are visually evident.
3. Incubate tissue in 1 ml of 0.05 % trypsin/EDTA for 5 min at 37 °C to dissociate. Dissociation reaction is terminated by the addition of an equal volume of trypsin inhibitor—mix well to ensure that trypsin is inactivated.
4. Centrifuge at 300 × *g* for 5 min and carefully aspirate the supernatant to remove all traces of trypsin.
5. Resuspend the cells in 250 μl of *Neurosphere Growth Medium*. Triturate cells using a P200 pipette tip to break up any remaining aggregated or clumped cells.
6. Add an additional 750 μl of *Neurosphere Growth Medium* and filter the cell suspension through a 40 μm mesh filter to remove tissue debris not digested by the trypsin.
7. Remove a 20 μl aliquot of the filtered cell suspension and mix with 80–180 μl Trypan Blue and count the number of viable cells using a hemocytometer (*see* **Note 1**).
8. Plate the filtered cell suspension (5,000 cells/cm^2) on uncoated tissue culture dishes in *Growth Medium* for 6–8 days. The cells are ready to be passaged when the majority of the Neurospheres are approximately 150 μm in diameter (*see* **Notes 2** and **3**).

3.2 Passaging of Neurosphere Cultures

When the Neurospheres are ready to be passaged, gently suspend the Neurospheres, and transfer the medium and suspended Neurospheres into a sterile conical tube.

1. Wash the culture dish one time with *Neurosphere Growth Medium* and pool the wash with the medium and cells in the conical tube.
2. Centrifuge at 100 × *g* for 5 min. Aspirate the supernatant and resuspend the cells in 1 ml of 0.05 % trypsin/EDTA.
3. Incubate at 37 °C for 2–3 min to dissociate the Neurosphere. Inactivate trypsin by adding an equal volume of trypsin inhibitor.

4. Centrifuge at 300 × *g* for 5 min. Aspirate the supernatant and resuspend the cells in 250 μl of *Neurosphere Growth Medium.*
5. Triturate cells 6–8 using a P200 pipette tip to make a single-cell suspension of cells.
6. Remove a 20 μl aliquot of the filtered cell suspension and mix with 80–180 μl Trypan Blue and count the number of viable cells using a hemocytometer (*see* **Note 4**).
7. Plate the cells in *Neurosphere Growth Medium* (5,000 cells/cm^2) on uncoated tissue culture and culture 6–8 days.

Neurospheres from passages 1–3 (P1–P3) are often irregularly shaped and disorganized clusters, with debris still present from the isolation procedure. After P3, healthy Neurosphere should be shiny, smooth with a uniform outline when viewed through a tissue culture microscope (*see* Fig. 1). When Neurospheres reach this state, they are ready for downstream studies.

The plating density of the neural stem cells in later passages is highly dependent on the purpose of the experiment. If the experimental outcome is to determine the clonality of isolated or treated cells, the cells must be plated at very low densities to ensure that the resulting Neurospheres are the result of single cell expansion and not the result of aggregation of multiple cells and smaller Neurospheres. The most stringent plating method calls for cells to be plated in 96-well plates at very low plating densities (1–10 cells/μl with 200 μl/well). The researcher can then identify which wells contain only a single cell and monitor just those wells for Neurosphere formation. If the experiment is designed to measure other outcomes such as the influence of growth factors or pharmacologic treatments on neuronal development, then less stringent plating conditions are required (*see* **Note 5**).

3.3 Differentiation of Neurospheres

3.3.1 Coat Chamber Slides or Coverslips with Poly-L-Ornithine/Laminin

1. Dilute the poly-L-ornithine stock solution 1:100 in cell culture-grade distilled water to make 10 μg/ml working solution.
2. Cover the surface of the chamber dish or coverslip with the 10 μg/ml poly-L-ornithine (typically about 1 ml for one well of a 2-well chamber slide or about 125–250 μl/cm^2) and incubate for 1 h at 37 °C.
3. Rinse the culture dish or coverslips twice with sterile water.
4. Cover the surface of the culture dish or coverslip with the laminin (10 μg/ml) and incubate for 2 h at 37 °C.
5. Rinse the culture dish or coverslips twice with PBS (*see* **Note 6**).

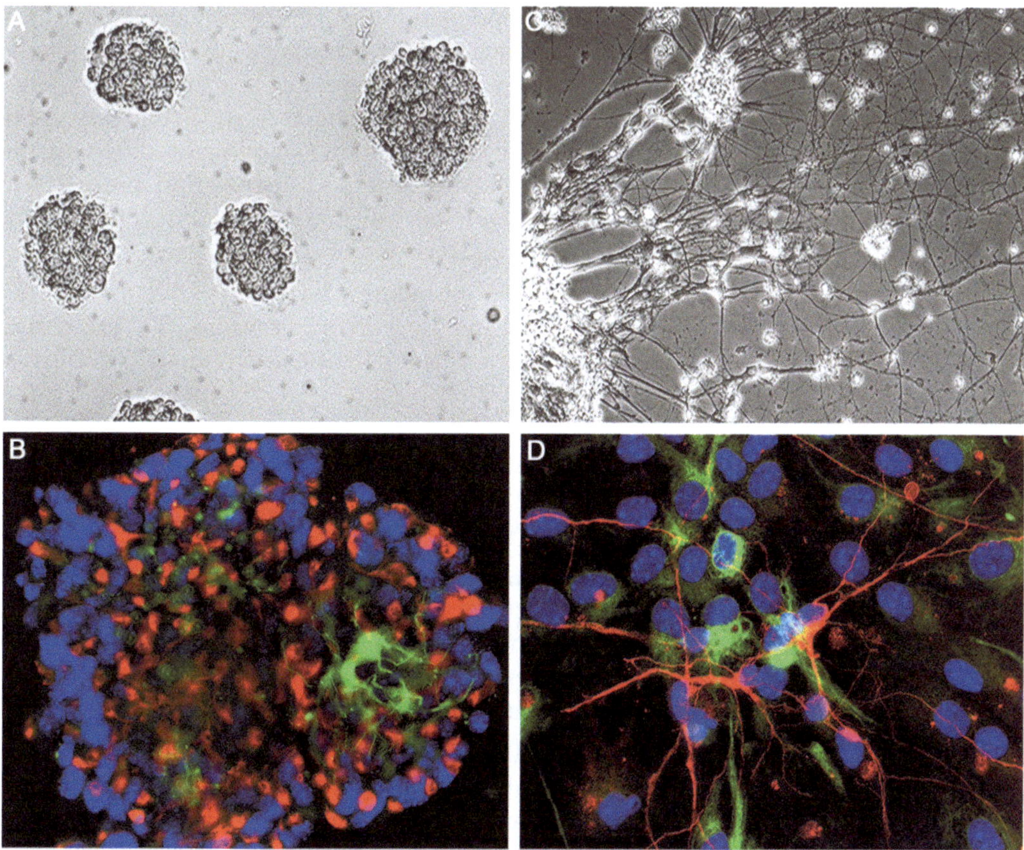

Fig. 1 Development and immunocytochemical characterization of Neurospheres from adult stem cells. (**a**) After 10 days in a cell growth promoting culture, Neural Stem Cell form floating spherical conglomerates. (**b**) Double labeling of these proliferating Neurospheres with Nestin (Rhodamine), and Class III β-Tubulin (Fluorescein) demonstrates that the cells are committed towards a neuronal lineage. While the majority of cells express the earlier marker Nestin, only few cells at this stage have started their Tubulin production. (**c**) After induction of differentiation for 5 days, a phase contrast micrograph of cultured Neurospheres show cells with slim and long processes that form a fibrillary net. (**d**) Double labeling for glial markers shows some large cells expressing GFAP (Rhodamine), mixed with a larger population of smaller, rounder cells that have differentiated into oligodendrocytes and express Gal-C (Fluorescein)

3.3.2 Differentiation of Intact Neurospheres

The Neurospheres are ready for plating when they reach approximately 150 nm in diameter (between day 6–8 of culture).

1. Collect the medium and Neurospheres from the culture dish and transfer to a conical tube and centrifuge at $100\times g$ for 5 min. Intact Neurospheres should be handled very gently in order to avoid dissociation.
2. Aspirate the *Growth medium* and gently resuspend the Neurosphere in *Differentiation medium*. Transfer the Neurospheres suspension to a sterile 60- or 100-mm^2 dish and select individual Neurospheres using a 1 ml pipette.

3. Transfer an appropriate number of Neurospheres to the tissue culture dishes containing the pre-coated coverslips or chamber slides. Neurospheres should be plated sparsely so that there is adequate space for the Neurospheres to attach and disperse on the coated surface without any merging of differentiating Neurosphere populations. The differentiation process typically takes 6–8 days, by which time the differentiated cells should form a monolayer on the culture surface.
4. Fix the cells (*see* Subheadings 3.4.1 and 3.4.2 for different fixation methods) and proceed to Subheading 3.4.3 to analyze the resulting population of differentiated cells by immunocytochemistry.

3.3.3 Differentiation of Dissociated Neurospheres

When Neurospheres are ready for harvesting, pre-coat glass coverslips or chamber slides and collect Neurospheres as described above (Subheadings 3.3.1–3.3.2).

1. Aspirate the *Neurosphere Growth medium* and resuspend the cells in 1 ml of 0.05 % trypsin/EDTA. Incubate at room temperature for 2–3 min. Add an equal volume of trypsin inhibitor and mix well to terminate the digestion. Do not over-trypsinize cells as this can lead to decreased viability of cells after plating.
2. Centrifuge at $300 \times g$ for 5 min and remove the supernatant. Immediately resuspend the cells in 250 μl of differentiation medium. Triturate the cells using a P200 pipette tip until only single cells are observed when the suspension is checked under the microscope.
3. Count the cells using a hemocytometer.
4. Dilute the appropriate volume of cell suspension in *Differentiation medium* and plate in tissue culture dishes containing pre-coated coverslips or in pre-coated chamber slides. There is a great deal of variability in the required plating density of dissociated cells. It is best to try a range of densities from 1×10^4 to 5×10^5 cell/cm^2 to determine which density works best for each experiment.
5. Differentiation should be complete after 4–7 days. Fix the cells (*see* Subheading 3.4.2 for different fixation methods) and proceed to Subheading 3.4 to analyze the differentiated cells by immunocytochemistry.

The resulting cell population will reflect the potentiality of the stem cell culture and can be used to test the effects of genetic or therapeutic manipulation on neural stem cell fate. Alternatively, culture conditions can be adjusted to selectively push neural stem cells into different lineages, as outlined below.

3.3.4 Culture Conditions to Preferentially Differentiate Neural Stem Cells Into Neurons

1. Follow the preceding steps to harvest and plate either intact or dissociated Neurospheres as described above (Subheadings 3.3.2 to 3.3.3).

2. Resuspend cells in *Neuronal Differentiation Medium* instead of the generic *Neurosphere Differentiation Medium.*
3. Culture the cells for 5–7 days. The medium can be replenished every 2 days by removing ½ to ¾ of the medium and replacing it with an equal volume of fresh *Neuronal Differentiation Medium* (*see* **Note 7**).
4. On Day 5–7, add 500 μM dibutyryl cAMP may be added to the differentiation medium and cells cultured for an additional 3 days to further select for neuronal cells.

3.3.5 Culture Conditions to Preferentially Differentiate Neural Stem Cells Into Astrocytes

1. Follow the preceding steps to harvest and plate either intact or dissociated Neurospheres as described above (Subheadings 3.3.2 to 3.3.3).
2. Resuspend cells in *Astrocyte Differentiation Medium* instead of the generic *Differentiation Medium.*
3. Culture the cells for 5–7 days. The medium can be replenished every 2 days.
4. On day 6–7, shake tissue culture dish overnight at 150 rpm in a 37 °C incubator remove non-adherent glial cells (oligodendrocytes) to further purify culture for astrocytes (*see* **Note 8**). Cells may be split and subjected to additional rounds of shaking when confluent to continue to select for astrocytic lineage.

3.3.6 Culture Conditions to Preferentially Differentiate Neural Stem Cells Into Oligodendrocytes

1. Follow the preceding steps to harvest and plate either intact or dissociated Neurospheres as described above (Subheadings 3.3.2 to 3.3.3).
2. Resuspend cells in *Oligodendrocyte Differentiation Medium* instead of the generic *Differentiation Medium.*
3. Culture the cells for 5–7 days. The medium can be replenished every 2–3 days.

4 Immunocytochemical Analysis of Cells Derived from Differentiated Neurospheres

4.1 Preparing Paraformaldehyde Fixing Solution

1. Prepare 20 % PFA stock solution by adding PBS to 20 g of PFA, and bring the volume up to 100 ml.
2. Add 250 μl 10 N NaOH, and heat the solution at 60 °C with stirring until the solution is completely dissolved.
3. Filter the solution through a 0.22-μm filter and cool on ice. Make sure that the pH is 7.5–8.0.
4. The solution can be aliquoted and stored at −20 °C for later use.
5. Before beginning immunocytochemistry protocol, dilute 20 % PFA to a working dilution of 4 % PFA using PBS (*see* **Note 9**).

4.2 Fixation and Permeablization of Differentiated Cells

1. Remove culture medium and gently wash the cells with PBS, while taking care not to dislodge the cells.
2. Fix cells with 4 % fresh PFA (as prepared above) at room temperature for 15 min.
3. Aspirate the PFA solution and wash three times with PBS, 5 min each wash (*see* **Notes 10** and **11**).
4. Permeabilize the cells with 0.2 % Triton X-100 for 5 min.
5. Aspirate the Triton solution and wash three times with PBS, 5 min each wash.

4.2.1 Fixation with Methanol

Fix the cells by adding ice-cold methanol. Incubate at −20 °C for 20 min. Aspirate methanol and wash three times with PBS.

4.2.2 Fixation with Acetone

Fix the cells by adding ice-cold acetone. Incubate at −20 °C for 5 min. Aspirate acetone and wash three times with PBS (*see* **Note 12**).

The choice of fixation method used will depend on both the epitope and primary antibody and therefore this step may require some optimization to determine which method is best. Fixation using cross-linking reagents, such as PFA are better at preserving cell structure, but may reduce the detection of some targets as cross-linking may interfere with antibody binding. Organic solvents like acetone or ethanol dehydrate and precipitate proteins on the cellular architecture and tend to work well for detection of cytoskeletal components but may reduce detection of soluble proteins.

4.3 Immunolabeling of Fixed Cells

1. Incubate cells for 1 h in blocking solution. The blocking solution should be a 5 % solution of serum derived from the secondary antibody host species diluted in PBS. For example, if the secondary antibody was generated in sheep, use 5 % normal sheep serum in PBS to block the slides/coverslips.
2. Remove the blocking solution and incubate the cells 2 h at RT or overnight at 4 °C with primary antibody diluted in 1 % serum. Ensure that the cell surfaces are covered uniformly with the antibody solution. If using coverslips, they can be placed on Parafilm in a humidified container and covered with antibody solution. This will reduce the volume of antibody required.
3. Wash the cells three times with PBS, 5 min each wash.
4. Incubate the cells with fluorescently labeled secondary antibody (diluted in 1 % serum in PBS) in the dark for 45–60 min at RT.
5. Wash the cells three times with PBS, for 5 min each wash.
6. Mount using ProLong® Gold anti-fade with DAPI mounting reagent and seal with the coverslip. Store the slides in the dark at 4 °C (*see* **Note 13**).

5 Immunocytochemical Analysis of Intact, Undifferentiated Neurospheres

The Neurospheres can be fixed and immunolabeled as intact Neurospheres. This may be desired if the goal is to identify pathways associated with maintenance of, or the differentiation from, the stem cell phenotype [7]. Immunocytochemical analysis of the intact Neurosphere would allow the investigation of how the expression of proteins of interest changes during the differentiation program, or to identify the presence of representative markers of apoptosis or cell cycle regulation following experimental manipulation of the neurosphere cultures. Identifying the patterns of protein expression in intact Neurospheres is advantageous since dissociation and plating of the Neural Stem cells initiates differentiation and alters gene expression. In addition, a model containing the majority of cells present in the brain allows one to track cytopathological changes to specific phenotypes, and to observe the interactions between different cell types.

5.1 Fixation and Permeablization of Intact Neurospheres

1. Thaw the Matrigel bottle at 4 °C overnight to prevent polymerization. The next day, prechill Neurobasal medium and 1.5 ml Eppendorf tubes. On ice, dilute Matrigel 1:50 with cold Neurobasal medium. Aliquot the undiluted Matrigel into chilled Eppendorf tubes and store at −20 °C.
2. Cover the surface of the chamber dish or coverslip with the diluted Matrigel and incubate for 2 h at 37 °C.
3. Collect the medium and Neurospheres from the culture dish and transfer to a conical tube and centrifuge at 100×*g* for 5 min. Intact Neurospheres should be handled very gently in order to avoid dissociation.
4. Aspirate the old *Growth medium* and gently resuspend the Neurospheres in fresh *Growth medium.* Transfer the Neurosphere suspension to a sterile 60- or 100-mm^2 dish and select individual Neurospheres using a 1 ml pipette.
5. Transfer an appropriate number of Neurospheres to the Matrigel-coated tissue culture dishes containing the precoated coverslips or chamber slides.
6. Incubate the Neurospheres at 37 °C for 20 min to allow the Neurospheres to attach to the Matrigel.
7. Fix in 4 % PFA for 15 min and wash one time with PBS for 20 min at room temperature.
8. Proceed with immunolabeling protocol discussed above (Subheading 4, **step 3**).

6 Neurospheres Generated fom Brain Tumor Stem Cells (BTSC)

Neurosphere assays can be utilized to identify neural stem-like properties in cells derived from primary brain tumors. Primary tumor cells have the propensity to retain their morphological and physiological feature when maintained in Neurosphere culture as compared to cells cultured in two dimensional cell cultures. Like Neural Stem Cells, Neurosphere-forming cells derived from primary tumors are self-renewing and retain the ability to differentiate into the cells found in the tumor in vivo [8, 9]. In addition, upon transplantation into nude mice, they form tumors that reflect the tumor cell composition from which they were derived [10–12]. The propensity of primary tumor-derived cells to form Neurospheres has been correlated with clinical outcome in both pediatric tumors and Glioblastomas [13]. The outcome of genetic manipulation and/or pharmacological treatments on BTSC could be analyzed using neurosphere assay. Immunocytochemical analysis of BTCS-derived Neurospheres could identify whether tumor stem cell markers are altered or preserved (*see* Fig. 2).

6.1 Coat Tissue Culture Dishes with Poly-HEMA

1. Before beginning, coat tissue culture dishes with poly-HEMA, which prevents attachment of tumor cells to tissue culture dish. To make a 20 mg/ml solution of poly-HEMA, dissolve 50 mg of poly-HEMA in 50 ml 95 % ethanol at 37 °C overnight.
2. Centrifuge at 2,000×*g*, 20 min to clarify the solution.
3. Cover culture vessel surface with 20 mg/ml poly-HEMA solution and allow to dry. Culture dishes may be sterilized under UV light and stored for 1 month.
4. Just before use, rinse culture dishes with sterile PBS.
5. Tumor cells are collected at time of surgical removal. Place immediately in ice-cold RPMI and keep on ice until they are processed for culturing for the Neurosphere assay.
6. Mince tissue using the blade of a scalpel for approximately 1 min, until only small pieces are evident.
7. Incubate tissue in 10 ml of Liberase/HBSS for 15–30 min at 37 °C to dissociate. Invert every 5 min to mix tissue with solution (*see* **Note 14**).
8. Once the cells are dissociated, mix the cell/enzyme reaction with a 1:1 volume of cold RPMI, 10 % FBS and mix by inversion. The presence of serum and the lower temperature inactivates the Liberase enzyme mix (*see* **Note 15**).
9. Centrifuge at 400×*g* for 5 min and carefully aspirate the supernatant.
10. Repeat **steps 4** and **5**.

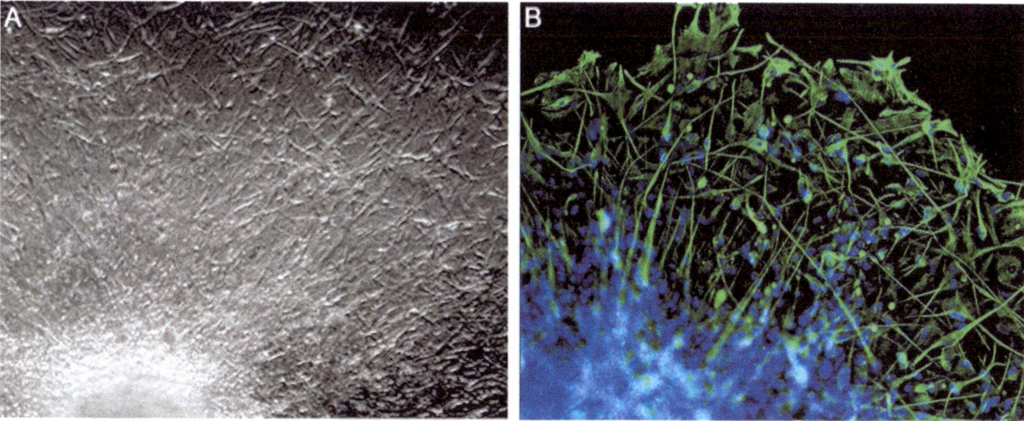

Fig. 2 Culture and Immunocytochemistry of three-dimensional cultures derived from primary glial tumors. (**a**) After dissociation from a primary diffuse fibrillary astrocytoma, a phase contrast micrograph of the resulting astrosphere shows the characteristic morphological features of astrocytic cells, with slim and long cytoplasmic processes. (**b**) Immunocytochemistry for Glial Fibrillary Acidic Protein (GFAP) corroborates the astrocytic phenotype of the cultured cells

11. Resuspend the cells in *Neurosphere Growth Medium*, and plate sparsely on poly-HEMA-coated dishes. Cells should be plated sparsely to avoid cell aggregation.
12. Tumor-derived Neurosphere should be handled as discussed for Neural Stem Cells-derived Neurospheres (Subheading 3.2).

7 Notes

1. The amount of Trypan Blue added will depend on the cell density of the cell suspension.
2. Avoid moving the Neurosphere-containing dishes too much to prevent mechanical aggregation of Neurospheres.
3. The Neurospheres should not be allowed to grow larger than 150 nm to preserve the viability and undifferentiated state of the inner cells.
4. If there is not a single-cell suspension, pass the cells through the P200 pipette tip several more times. It is important the cells are plated as individual cells to help ensure that the resultant Neurospheres are the result of clonal expansion and do not arise from multiple cells.
5. Cells grow better if medium is changed ½ a time.
6. Coverslips or culture dishes may be prepared in advance. Keep the coated surfaces covered with PBS, wrapped tightly with Parafilm, at RT for up to a week.

7. Neurons are not particularly adherent. Be careful not to dislodge them during media changes.
8. These floating cells can be taken and re-plated in Oligodendrocyte Diff Medium to concurrently generate an oligodendrocyte culture.
9. It is best to use freshly diluted 4 % PFA.
10. If cells are easily dislodged/washed off, they may be washed just one time with a greater volume of PBS.
11. Fixed cells may be stored for up to 3 weeks in PBS at 4 °C before staining. Do not allow slides to dry.
12. If methanol or acetone is used as a fixatives, a permeabilization step is not needed.
13. Include a secondary only control (in Subheading 3.4.3 omit **step 2**) to ensure that observed signal is specific to primary antibody. If co-labeling, also label slides with singly with each primary antibody.
14. Can be 10–25 ml depending on the starting weight of the tumor sample.
15. The cell suspension may be filtered through a 40 μm mesh filter to remove debris.

References

1. Eriksson PS, Perfilieva E, Björk-Eriksson T et al (1998) Neurogenesis in the adult human hippocampus. Nat Med 4:1313–1317
2. Lois C, Alvarez-Buylla A (1993) Proliferating subventricular zone cells in the adult mammalian forebrain can differentiate into neurons and glia. Proc Natl Acad Sci USA 90: 2074–2077
3. Doetsch F, Alvarez-Buylla A (1996) Network of tangential pathways for neuronal migration in adult mammalian brain. Proc Natl Acad Sci USA 93:14895–14900
4. Kirschenbaum B, Nedergaard M, Preuss A et al (1994) In vitro neuronal production and differentiation by precursor cells derived from the adult human forebrain. Cereb Cortex 4:576–589
5. Reynolds BA, Weiss S (1992) Generation of neurons and astrocytes from isolated cells of the adult mammalian central nervous system. Science 255:1707–1710
6. Rietze RL, Reynolds BA (2006) Neural stem cell isolation and characterization. Methods Enzymol 419:3–23
7. Nyfeler Y, Kirch RD, Mantei N et al (2005) Jagged1 signals in the postnatal subventricular zone are required for neural stem cell self-renewal. EMBO J 24:3504–3515
8. Hemmati HD, Nakano I, Lazareff JA et al (2003) Cancerous stem cells can arise from pediatric brain tumors. Proc Natl Acad Sci USA 100:15178–15183
9. Vescovi AL, Galli R, Reynolds BA (2006) Brain tumour stem cells. Nat Rev Cancer 6: 425–436
10. Singh SK, Clarke ID, Terasaki M et al (2003) Identification of a cancer stem cell in human brain tumors. Cancer Res 63:5821–5828
11. Galli R, Binda E, Orfanelli U et al (2004) Isolation and characterization of tumorigenic, stem-like neural precursors from human glioblastoma. Cancer Res 64:7011–7021
12. Lee J, Kotliarova S, Kotliarov Y et al (2006) Tumor stem cells derived from glioblastomas cultured in bFGF and EGF more closely mirror the phenotype and genotype of primary tumors than do serum-cultured cell lines. Cancer Cell 9:391–403
13. Laks DR, Masterman-Smith M, Visnyei K et al (2009) Neurosphere formation is an independent predictor of clinical outcome in malignant glioma. Stem Cells 27(4):980–987

Chapter 11

Transfection of Neuronal Cultures

Ilker Kudret Sariyer

Abstract

Efficient transfection of genes into neurons is a crucial step for the study of neuronal cell biology and functions. These include but are not limited to investigating gene function by overexpression of target proteins via expression plasmids and knocking down the expression levels of neuronal genes by RNA interference (RNAi). In addition, reporter gene constructs are widely used to investigate the promoter activities of neuronal genes. Numerous transfection techniques have been established to deliver genes into the cells. However, efficient transfection of post-mitotic cells, including neurons, still remains a challenging task. Here, we overview the advantages and disadvantages of various techniques for the transfection of primary neurons, and provide an optimized protocol for FuGENE-6 (Promega) which allows a suitable transfection efficiency of primary neuronal cultures.

Key words Primary neurons, Gene delivery, Transfection, Electroporation, Nucleofection, Calcium co-precipitation, Lipofection, FuGENE-6

1 Introduction

Transfection is the process of introducing nucleic acids into cells. The term transfection is commonly used for nonviral gene delivery [1, 2]. Transfection of animal cells typically involves opening transient pores in the cellular membrane that will allow the uptake of genetic material into the cytoplasm. Transfection may cause morphologic changes and abnormalities in target cells. The transfection of primary cortical and hippocampal neurons has proven to be a real challenge for the study of neuronal gene functions [2, 3]. Over the past three decades, many transfection methods have been developed including electroporation, nucleofection, calcium phosphate co-precipitation, and lipofection.

Among these transfection methods, electroporation alters the phospholipid bilayer's hydrophobic/hydrophilic interactions of plasma membrane by exposing cells to a voltage pulse [3–5]. Applying a quick voltage shock disrupts areas of the membrane temporarily, allowing polar molecules to enter the cells through the membrane. Electroporation is generally used with freshly

Shohreh Amini and Martyn K. White (eds.), *Neuronal Cell Culture: Methods and Protocols*, Methods in Molecular Biology, vol. 1078, DOI 10.1007/978-1-62703-640-5_11, © Springer Science+Business Media New York 2013

isolated neurons or neuronal cell lines in suspension, and requires specialized equipment which is relatively expensive. In order to avoid high cellular mortality associated with electroporation, the nucleofection technique has been developed. In conventional electroporation, the cell membrane is broken down by electric pulses and DNA from the surrounding buffer enters the cytoplasm resulting with high cell mortality. To overcome this problem, nucleofection applies cell-type-specific combinations of electric currents and solutions and thus increases the cell survival [2, 6, 7]. Although nucleofection provides high levels of transfection efficiency with low toxicity, its application is also restricted to neuronal progenitors and freshly isolated neurons in suspension.

The calcium phosphate co-precipitation method is one of the most widely used transfection methods of primary neuronal cultures [8–10]. The principle of the technique relies on formation of DNA crystals with Ca^{2+} ions in the phosphate buffer. The crystals precipitate onto the cells during the transfection, and enter the cells by endocytosis. The technique does not require any specialized equipment. It is cost-effective and easy to optimize for the best results. Although the method is suitable to transfect neurons at all stages of differentiation, in postmitotic neurons, delivery of the DNA into the nucleus is more difficult and that leads to low transfection efficiencies [11].

Lipofection is perhaps the most popular transfection method for the gene delivery into a variety of animal cells. The technique relies on cationic lipid molecules which form unilamellar liposomes [12–15]. The positively charged (cationic) lipid molecules interact with negatively charged DNA to form a DNA–lipid aggregate [12, 13]. Genetic material is delivered into cells by this lipid aggregate which fuses with the negatively charged plasma membrane. In order to increase the fusion capacity, cationic lipid molecules are often combined with neutral helper lipid molecules. Many different formulations of lipofection reagents have been developed. The lipofection technique is easy to optimize and generally less toxic to the cells. Although the technique produces high transfection efficiencies in a wide range of cell types, the transfection efficiency of postmitotic neurons shows variable results based on the lipofection reagent used. In order to determine the best suitable transfection method for a particular experiment, the specific advantages and disadvantages of each technique must be considered. Each method has been proven to have their own strengths and drawbacks regarding toxicity, expression levels, cell viability, and transfection efficiency [3]. In Table 1, the advantages and disadvantages of commonly used transfection techniques for primary neurons is summarized. Below we provide an optimized protocol of FuGENE-6 (Promega) which we have developed in our laboratory and yields relatively high transfection efficiencies in primary neuronal cultures.

Table 1
Transfection methods commonly used to transfect primary neurons: advantages and disadvantages

	Advantages	Disadvantages
Electroporation	Relatively high transfection efficiencies Less optimization required Simple and quick transfection Expression of the genes starts within hours	Only used for freshly isolated neurons in suspension Expensive material Relatively higher toxicity Requires specialized equipment
Nucleofection	Delivers genes directly into the nucleus Very high transfection efficiencies Higher expression levels Requires specialized equipment Relatively low toxicity	Only used for freshly isolated neurons in suspension Requires expensive material and equipment Requires optimization of the material
Ca2+ phosphate Co-precipitation	Can be used for cultured primary neurons No specialized equipment is required Little optimization for different plasmids Low toxicity when optimized Cost-effective	Relatively low transfection efficiencies
Lipofection	Can be used for cultured primary neurons Simple and quick transfection procedure Provides reproducible transfection efficiencies High transfection efficiency depending on reagent Cost-effective	Needs optimization for higher transfection efficiencies Depending on reagent used, some toxicity is observed

2 Materials

1. Freshly isolated or post-mitotic neurons.
2. 6-well cell culture-grade plastic dishes.
3. Neuronal Growth Medium (NGM): Prepare Neuro-Basal Medium supplemented with B27 (2 %), L-Glutamine (250 μM), Penicillin/Streptomycin (50 μg/ml), and amphotericin b (2.5 μg/ml).
4. Opti MEM serum free media (Life Technologies).
5. FuGENE-6 transfection reagent (Promega).
6. Plastic pipets.
7. pLEGFP-C1 plasmid.
8. 37 °C, 5 % CO_2 humidified tissue culture incubator.
9. Sterile tissue culture hood.
10. Sterile 1.5 ml microcentrifuge tubes.

3 Methods

1. Plate and maintain neurons in 6-well tissue culture dishes at a density of 60–80 % confluency (*see* **Note 1**).
2. Place Opti-MEM serum free media at 37 °C for at least 1 h prior transfection. Opti-MEM is provided in 500 ml bottles. To maintain the quality and freshness of the media, aliquot cold Opti-MEM in 15 or 50 ml sterile tubes based on amount to be needed for the transfection.
3. FuGENE6 is stored at 4 °C. Bring FuGENE6 at room temperature before usage.
4. Transfection mixture should be prepared in tissue culture hood where contamination risk is minimized. Prepare and label sterile 1.5 ml microcentrifuge tubes based on transfection conditions.
5. Place 100 μl of Opti-MEM media into the 1.5 ml microcentrifuge tubes, add 18 μl FuGENE6 to form transfection mixture A (TM-A). Add the transfection reagent directly to the Opti-MEM. Put the tip directly into the Opti-MEM do not allow the reagent to touch any other plastic parts of the tube. FuGENE-6 may bind to the plastic and limit the effectiveness of the transfection (*see* **Notes 2–4**).
6. Mix the TM-A by gentle pipetting. Do not vortex or centrifuge the TM-A mixture.
7. In a separate tube, prepare transfection mixture B (TM-B). Dilute DNA construct (pLEGFP-C1 plasmid) in 100 μl of Opti-MEM media as 60 ng/μl (total DNA amount for each well is 6 μg).
8. Incubate TM-A and TM-B in the cell culture hood for 5 min, and carefully combine TM-B with TM-A in the TM-A tube to obtain primary transfection mixture (PTM). Mix the PTM by gentle pipetting. Do not vortex or centrifuge the PTM mixture.
9. Incubate PTM at room temperature for 30 min. This will allow proper time for the formation of DNA–liposome complexes.
10. Remove media from primary neuronal cultures, and rinse the cells with 2 ml warm Opti-MEM.
11. The volume of PTM is about ~220 μl (this will change slightly based on DNA amount). Add 280 μl more Opti-MEM media to bring the volume to the total of 500 μl (final PTM volume).
12. Add the PTM mixture directly (slowly and drop-wise) onto the cells while swirling media in cells to mix.

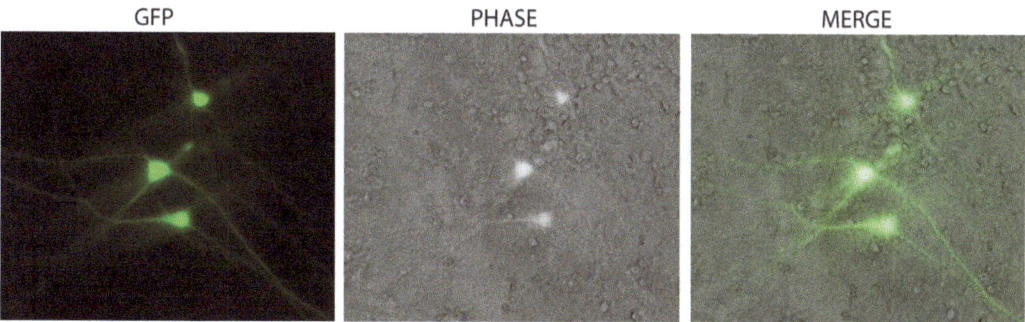

Fig. 1 GFP expression in postmitotic neurons 72 h after transfection with FuGENE-6. Primary neurons were plated in 6-well tissue culture plates at a density of 60 % confluency and transfected with pLEGFP-C1 plasmid as described in methods section

13. Place cells back into the tissue culture incubator and incubate for 6 h. It is important to swirl the transfection media on cells every hour during the transfection process. That will help to yield higher transfection efficiencies (*see* **Note 5**).
14. After 6 h of incubation, remove the transfection mixture, and add 2 ml fresh growth media onto cells.
15. Green fluorescein protein (GFP) expression will typically start around 6 h post-transfection, and will reach to a pick at 48–72 h (*see* Fig. 1).

4 Notes

1. Isolation and preparation of neuronal cultures for the transfection is a crucial step (use appropriate protocols for the isolation of cortical and hippocampal neurons). Neurons should be isolated and plated in sterile tissue culture plates at a density of 50–80 % and maintained in a humidified tissue culture incubator. Using too low a cell density may cause cells to grow poorly. If the cultures are too dense, this may result in contact inhibition which negatively affects the uptake of DNA. Cultures should be free of contamination, and grown in appropriate medium with all the necessary growth factors.
2. Use high-quality of DNA which is free of RNA, proteins, and endotoxins. The DNA should be of transfection-quality. This can be achieved by using a high quality plasmid preparation kit such as PureYield Plasmid Maxiprep Systems. The A_{260}–A_{280} ratio of the DNA should be 1.7–1.9.
3. The optimal amount of DNA is crucially important. Determining the right amount of the plasmid DNA for transfection is important to obtain high transfection efficiencies. The amount of DNA will vary widely depending upon the

Table 2
Reference table for PTM composition for different size of tissue culture plates

	Growth area (cm^2)	Total DNA amount (μg)	PT-A volume (μl)	PT-B volume (μl)	FuGENE-6 volume (μl)	Final PTM volume (μl)
96-well plate	0.32	0.375	6	6	1.2	24
24-well plate	2.0	1.5	25	25	4.5	125
12-well plate	3.8	3.0	50	50	19	250
6-well plate	9.6	6.0	100	100	18	500
60 mm dishes	21.3	12	200	200	36	1,000
100 mm dishes	58.1	20	400	400	60	2,000

transfection reagent, DNA type, number of cells, and the surface area of the tissue culture dishes. Table 2 summarizes the optimized composition of PTM used to transfect primary neurons via FuGENE-6 in different size of tissue culture plates.

4. Attaining an optimal ratio of FuGENE-6 to the DNA is also very important. Ratios of 3:1 and 2:1 work well for primary neurons, but ratios outside this range may be optimal for a particular experiment or application.
5. Toxicity is an important consideration. FuGENE-6 transfection reagent is one of the less toxic methods of DNA transfection into cells. In the event of cell death, optimize conditions as follows. Lower the amount of input DNA and FuGENE-6 while holding the DNA–FuGENE-6 ratio constant. Increasing the cell density on the plates may also eliminate the associated toxicity. In the case of excessive cell death, possible toxicity of the gene products should also be considered. That can be addressed by using a control plasmid for transfections performed in parallel.

References

1. http://www.promega.com/resources/product-guides-and-selectors/protocols-and-applications-guide/transfection/
2. Zeitelhofer M, John P, Vessey JP, Thomas S, Kiebler M, Dahm R (2009) Transfection of cultured primary neurons via nucleofection. Curr Protoc Neurosci 47:4.32.1–4.32.21
3. Karra D, Dahm R (2010) Transfection techniques for neuronal cells. J Neurosci 30(18):6171–6177
4. Washbourne P, McAllister AK (2002) Techniques for gene transfer into neurons. Curr Opin Neurobiol 12:566–573
5. Dib-Hajj SD, Choi JS, Macala LJ, Tyrrell L, Black JA, Cummins TR, Waxman SG (2009) Transfection of rat or mouse neurons by biolistics or electroporation. Nat Protoc 4:1118–1126
6. Gresch O, Engel FB, Nesic D, Tran TT, England HM, Hickman ES, Korner I, Gan L, Chen S, Castro-Obregon S, Hammermann R, Wolf J, Muller-Hartmann H, Nix M, Siebenkotten G, Kraus G, Lun K (2004) New non-viral method for gene transfer into primary cells. Methods 33:151–163

7. Gartner A, Collin L, Lalli G (2006) Nucleofection of primary neurons. Methods Enzymol 406:374–388
8. Chen C, Okayama H (1987) High-efficiency transformation of mammalian cells by plasmid DNA. Mol Cell Biol 7:2745–2752
9. Yu LY, Arumäe U (2008) Survival assay of transiently transfected dopaminergic neurons. J Neurosci Methods 169(1):8–15
10. Gabellini N, Minozzi MC, Leon A, Dal TR (1992) Nerve growth factor transcriptional control of c-fos promoter transfected in cultured spinal sensory neurons. J Cell Biol 118(1):131–138
11. Goetze B, Grunewald B, Baldassa S, Kiebler M (2004) Chemically controlled formation of a DNA/calcium phosphate coprecipitate: application for transfection of mature hippocampal neurons. J Neurobiol 60:517–525
12. Felgner PL, Gadek TR, Holm M, Roman R, Chan HW, Wenz M, Northrop JP, Ringold GM, Danielsen M (1987) Lipofection: a highly efficient, lipid-mediated DNA-transfection procedure. Proc Natl Acad Sci USA 84:7413–7417
13. Felgner JH, Kumar R, Sridhar CN, Wheeler CJ, Tsai YJ, Border R, Ramsey P, Martin M, Felgner PL (1994) Enhanced gene delivery and mechanism studies with a novel series of cationic lipid formulations. J Biol Chem 269(4):2550–2561
14. Wiesenhofer B, Humpel C (2000) Lipid-mediated gene transfer into primary neurons using fugene: comparison to C6 glioma cells and primary glia. Exp Neurol 164:38–44
15. Ohki EC, Tilkins ML, Ciccarone VC, Price PJ (2001) Improving the transfection efficiency of post-mitotic neurons. J Neurosci Methods 112:95–99

Chapter 12

Lentiviral Transduction of Neuronal Cells

Hassen S. Wollebo, Baheru Woldemichaele, and Martyn K. White

Abstract

Here we describe a general method for the construction of a lentivirus vector using a specific example of the construction of a lentivirus containing the luciferase reporter gene under the control of two hypothetical promoters and derived HIV-1 based lentivirus expression vector pLVX-Puro. This method can be used to compare the strength and regulation of different promoters. In this example, the target cells for transduction are human primary fetal astrocytes but the method is applicable to any primary cell culture from the CNS or other tissue and can be used to examine the strength of a particular promoter in different cell types. HIV based lentivirus particles are prepared by transfection of 4 plasmids into 293T cells using the Fugene 6 transfection reagent.

Key words Lentiviral vectors, Transduction, Primary human fetal astrocytes

1 Introduction

Lentiviruses are a complex genus of retroviruses, which were named from "lente," which is the Latin word for slow referring to the prototypic slow progressing neurologic disease of sheep caused by maedi/visna virus [1]. They are enveloped particles containing homodimers of the linear single-stranded RNA genome [2]. Lentiviruses resemble γ-retroviral vectors and stably integrate into genome of the host cell, allowing the persistent expression of the gene of interest. However, unlike retroviruses, lentiviruses require active transport of the pre-integration complex across the nuclear pore by the action of the nuclear import machinery of the host cell [3], which allows lentiviruses to infect both dividing and nondividing cells [4–6].

The first lentiviral vectors were derived from HIV [7], the most extensively studied lentivirus. The incorporation of envelope proteins of other viruses in HIV particles was a significant step for the development of HIV based vectors. Most of HIV derived lentiviruses contain the envelop protein of rhabdovirus family (VSVG) instead of the HIV envelope glycoprotein. This modification not only increase host-range of VSVG pseudotyped lentiviruses but also increases the ability to concentrate VSV-G pseudotyped

Shohreh Amini and Martyn K. White (eds.), *Neuronal Cell Culture: Methods and Protocols*, Methods in Molecular Biology, vol. 1078, DOI 10.1007/978-1-62703-640-5_12,

particles more than 1,000-fold by ultracentrifugation. Lentiviral vectors (LVS) have become one of the most widely used vectors for the fundamental biological research. Currently, most lentiviral vectors are produced using transient transfection of packaging and vector plasmids. Using the 293T cells, which are susceptible to efficient transfection, one routinely obtains titers of 1×10^9 to 1×10^{10} IU/ml following transient transfection with the most recent generation of packaging and vector constructs, and then concentrating of the virus particles using ultracentrifugation [3].

Lentiviral vectors have been successfully used to transduce most cell types within the central nervous system (CNS) in vivo, including neurons, astrocytes, adult neuronal stem cells, oligodendrocytes, and astrocytes. Lentiviruses are also well suited for the transduction of various stem/progenitor cells, including hematopoietic stem cells (HSC) [5, 8, 9].

It has been reported that the high level expression of the gene of interest in lentivirus transduction of different cell lines in CNS depends on the strength of the promoter used. If the promoter is active in glia cells, then high level transgene expression in astrocytes can be achieved. The activity of the internal promoter used in lentiviral transduction determines the different expression pattern that have been reported although it is likely that the different cell types in the CNS can be transduced with lentivirus with similar efficiencies (*see* Fig. 1) [9].

2 Materials

1. Complete DMEM medium: 4 mM L-glutamine; 4.5 g/l D-GLUCOSE; 100 U/ml Penicillin–Streptomycin; 10 % heat-inactivated FBS (*see* **Note 1**).
2. 293T cells: 293T cells are good for viral production. It is essential that the cells be well maintained and of relatively low passage number. 293T Cells were maintained in a complete DMEM.
3. Primary fetal astrocytes: Purified primary human fetal astrocytes (PHFA) will be prepared as we have previously described [10] using a method based on Cole and de Vellis [11], and Yong and Antel [12]. Briefly, 12–16-week-old human fetal brain tissue is obtained from Advanced Biosciences Resources, Inc. (Alameda, CA) and dissected to remove blood vessels and meninges. Tissue is mechanically disrupted and digested with trypsin and DNase I. Cells are passed through a 70 μm filter to remove debris and pelleted by centrifugation. Mixed primary cultures are grown for 5 days in culture and then astrocytes purified. This is done by briefly trypsinizing cultures to remove non-astrocytic cells and washing the astrocyte monolayers. Culture purity is ascertained by staining for glial fibrillary acidic protein.

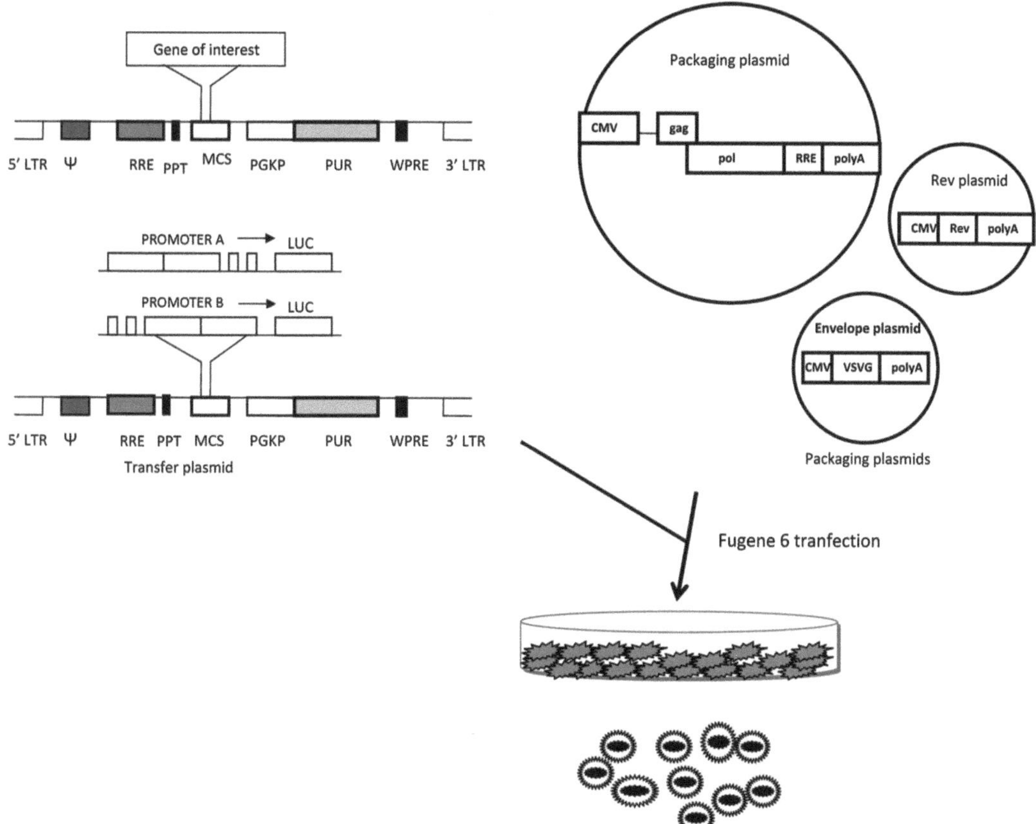

Fig. 1 Production of transducing lentivirus. Diagram illustrating the transduction of the target cells using the packaging plasmids and transfer plasmid. The transfer plasmid contains the gene of interest (in this case luciferase driven by promoter A or B), the psi packaging element (ψ), a polypurine tract DNA sequence (PPT) to increase the transduction efficiency, and *cis*-acting transcriptional regulatory elements, rev response element (RRE), the woodchuck hepatitis posttranscriptional regulatory element (WPRE) to improve the transgene expression, and the Puromycin resistance gene (Pur) driven by the phosphoglycerate kinase promoter (PGKP). The REV protein, which is important for the expression of gag and pol, is provided in *trans* with a separate plasmid. The viral stocks produced by the transfected cells can be used to transduce parallel cultures of target cells and compare the strength and regulation of promoters A and B

4. Plasmids: The purity of the DNA is important for transfection. Plasmids should be prepared using a high quality kit such as that from Qiagen. Plasmid systems are used for the production of lentivirus: pLVX-Puro is an HIV-1 based lentiviral expression vector with gene of interest and three lentiviral packaging plasmids pMDlg/PRRE(expresses HIV gag/pol), Prsv-REV(express HIV REV), and pMD2.G(express VSV glycoprotein) (*see* **Note 2**). These packaging plasmids can be purchased from Clontech.
5. Penicillin–Streptomycin (100×) solution of 10,000 U/ml Penicillin G sodium and 10,000 μg/ml streptomycin sulfate.
6. TrypLE express (GIBCO) (*see* **Note 3**).

7. Dulbecco's phosphate buffered saline (DPBS).
8. High-efficiency Transfection reagent (e.g., Fugene 6, Promega).
9. High-efficiency Plasmid Maxiprep Kit (e.g., Qiagen).
10. Cell freezing medium with 5 % DMSO, 20 % FBS.
11. Cryovials for freezing virus stocks.

3 Methods

3.1 Lentiviral Transfection and Production

1. Plate 4×10^6 293T cells in a T-75 flask 24 h before transfection. It is important that cells be well maintained and of relatively low passage number.
2. Transfect cells with a total of 8 μg DNA.

   In a 15 ml polystyrene culture tube, mix the following

   - 500 μl Opti-MEM medium.
   - 3 μl/μg DNA Fugene 6 reagent.

   In another 15 ml polystyrene culture tube prepare the DNA mix

   - 100 μl Opti-MEM.
   - 3 μg of pLVX-Puro vector.
   - 1.5 μg VSVG.
   - 1.2 μg RSV-REV.
   - 2.3 μg pMDL g/p RRE.
3. Incubate the two tubes 5 min separately.
4. Add the content of tube number 2 to tube number 1 and incubate the mixture for additional 25 min (*see* **Note 4**).
5. By the end of the above incubation time, prepare in 50 ml falcon tube 5.6 ml of DMEM and add the Fugene DNA mix directly to this medium and mix well. Take 293T cells out of the incubator and remove the medium from the cells and wash the cells once with PBS and add transfection mixture to the cells distribute evenly by swirling the flask. Return the plates immediately to incubator.
6. After 4 h, add an additional 4 ml fresh medium to the flask and incubate overnight.
7. The next day remove the medium and wash the cells 2× with PBS, and add 10 ml of fresh medium containing 6 μg/ml Polybrene.
8. After 48 h from the first transfection, harvest the lentivirus containing supernatants and refeed the cells with new fresh medium.

9. Centrifuge lentivirus containing supernatants 10 min at 787 × *g* at 4 °C in 50 ml tube to remove debris and filter viral supernatant through 0.45 μm filter and keep it at 4 °C for 1 day (*see* **Note 5**).
10. On the next day, collect the second round of the supernatant from the cells and process it in the same way as the first supernatant.
11. Combine the two supernatant together and concentrate by ultracentrifugation using SW-28 rotor at 140,000 × *g* for 90 min at 4 °C.
12. Aspirate the supernatant carefully and resuspend pellet in DMEM and leave tube at 4 °C overnight with gently shaking.
13. Centrifuge the viral suspension for 5 min at 7,580 × *g*, 4 °C and carefully remove the supernatant that contain the concentrated viral particles.
14. Virus particles should be aliquoted, flash frozen in liquid nitrogen and stored at −80 °C. Avoid multiple freezing and thawing which leads to reduction on vector titers (*see* **Note 6**).

3.2 Titration of Lentivirus

We determine viral titer by infecting primary astrocytes with serial dilution of concentrated virus. One day after transduction, the cells were harvested and luciferase activity of measured as indication of the efficiency of infection.

Transduction of primary fetal astrocytes

1. One day before transduction, plate primary astrocytes (in duplicate) in 6 well plate at cell density of 6×10^5 cells/well.
2. On the next day, aspirate the medium and wash the cell once with PBS and incubate the cell with 2 ml of viral particles containing DMEM in the presence of 6 μg/ml polybrene.
3. Incubate the cells overnight.
4. Next day remove the medium and replace with fresh medium and incubate the cells for 24 h before measuring promoter activity.

3.3 Conclusion

In this protocol, we described a simple and reproducible method of lentiviral vector production in 293 cells using the four plasmids system based on the Fugene 6 transfection method. Using this method of transfection significantly reduced the amount of DNA needed for transfection compared to the traditional method of calcium phosphate precipitation. The Fugene 6 transfection reagent used in the protocol can also be left on the cells for 24 h or longer without having a toxic effect. Finally, we recommend the use of a transfer plasmid with a reporter gene like GFP that significantly facilitates the determination of the transduction efficiency and viral titer.

4 Notes

1. It is essential that the cells be well maintained and of relatively low passage number. The proper cell density is crucial for high production yield.
2. The purity of the DNA for transfection is very critical as is the establishment of the optimal ratio of packaging, envelope and vector plasmids and the amount of Fugene 6 transfection reagent.
3. TrypLE™ Express is a special formulation of Trypsin/EDTA optimized for the culture of neurons.
4. Prolonged incubation of DNA and Fugene 6 mix decreases the transfection efficiency.
5. To filter the virus supernatant use filter made of cellulose acetate or low protein binding filters. It has been reported that filters with Nitrocellulose inactivate or destroy lentivirus particles by binding to the envelope.
6. Aliquot and store your virus stock in such way that multiple freezing–thawing step will be avoided. Frequent freezing and thawing steps lead to significant loss to viral titer.

References

1. Pfeifer A, Verma IM (2001) Gene therapy: promises and problems. Annu Rev Genomics Hum Genet 2:177–211
2. Denning W, Das S, Guo S et al (2013) Optimization of the transductional efficiency of lentiviral vectors: effect of sera and polycations. Mol Biotechnol 53(3):308–314
3. Mátrai J, Chuah MK, VandenDriessche T (2010) Recent advances in lentiviral vector development and application. Mol Ther 18:477–490
4. Sakuma T, Barry MA, Ikeda Y (2012) Lentiviral vectors: basic to translation. Biochem J 443:603–618
5. van Hooijdonk LW, Ichwan M, Dijkmans TF et al (2009) Lentivirus-mediated transgene delivery to the hippocampus reveals sub-field specific differences in expression. BMC Neurosci 10:2
6. Naldini L, Blomer U, Gage FH et al (1996) Efficient transfer, integration, and sustained long-term expression of the transgene in adult rat brain injected with a lentiviral vector. Proc Natl Acad Sci U S A 93:11382–11388
7. Naldini L, Blomer U, Gallay P et al (1996) In vivo gene delivery and stable transduction of nondividing cells by a lentiviral vector. Science 272:263–267
8. Logan AC, Lutzko C, Kohn B (2002) Advances in lentiviral vector design for gene-modification of hematopoietic stem cells. Curr Opin Biotechnol 13:429–436
9. Varma NR, Janic B, Ali MM et al (2011) Lentiviral Based Gene transduction and promoter studies in Human Hematopoietic Stem Cells (hHSCs). J Stem Cells Regen Med 7:41–53
10. Radhakrishnan S, Otte J, Enam S et al (2003) JC virus-induced changes in cellular gene expression in primary human astrocytes. J Virol 77:10637–10644
11. Cole R, de Vellis J (1997) Astrocyte and oligodendrocyte culture. In: Fedoroff S, Richardson A (eds) Protocols for neural cell culture. Humana, Totowa, NJ
12. Yong VW, Antel JP (1997) Culture of glial cells from human brain biopsies. In: Fedoroff S, Richardson A (eds) Protocol for neural cell culture. Humana, Totowa, NJ

Chapter 13

Compartmentalized Neuronal Cultures

Armine Darbinyan, Paul Pozniak, Nune Darbinian, Martyn K. White, and Kamel Khalili

Abstract

The culturing of neurons results in formation of the layer of neurons with random extensive overlapping outgrowth. To understand specific roles of somas, axons, and dendrites in complex function of neurons and to identify molecular mechanisms of biological processes in these cellular compartments various methods were developed. We utilized *AX*on *I*nvestigation *S*ystem (AXIS™) manufactured by Millipore. This device provides an opportunity to orient neuronal outgrowth and spatially isolate neuronal processes from neuronal bodies. AXIS device is a slide-mounted microfluidic system, which consists of four wells. Two of the wells are connected by a channel on each side of the device. Channels are connected by microgrooves (approximately 120). The size of microgrooves (10 μm in width and 5 μm in height) do not permit passage of cell through while allowing extension of neurites. The microfluidic design also allows an establishment of a hydrostatic gradient when the volume in one chamber is greater than that in the other (Park et al. Nat Protocols 1: 2128–2136, 2006). This feature allows studying of the effect of specific compounds on selected compartments of neurons.

Key words Compartment cultures, Neurite outgrowth, Axonal processes

1 Introduction

One of the major characteristics of neurons is that they possess axons and dendrites, which mediate the polarized intercellular transmission of information. Thus culture systems that permit analysis of these processes are highly valuable. Two compartment systems have now been developed that allow the investigation of vectoral functions in neuronal cell cultures, and in this chapter, we describe the *AX*on *I*nvestigation *S*ystem (AXIS™) device. This device allows the opportunity to orient neuronal outgrowth and spatially isolate neuronal processes from neuronal bodies using wells connected by channels with microgrooves that are small enough to prevent the passage of the cell body but large enough to allow extension of neurites [1]. This device can be used to perform a variety of experiments on the control of the extension of neurites, intercellular communication, retrograde and anterograde

Shohreh Amini and Martyn K. White (eds.), *Neuronal Cell Culture: Methods and Protocols*, Methods in Molecular Biology, vol. 1078, DOI 10.1007/978-1-62703-640-5_13, © Springer Science+Business Media New York 2013

transport and the isolation of the specific constituents of neurites, such as mRNAs for RT-PCR, apart from those of the soma of the neuronal cell cultures.

2 Materials

Isolate human fetal cortical neurons according the protocols described in Chapter 2 (**steps 1–6**). Carry out all procedures in sterile conditions at room temperature unless otherwise specified.

2.1 Brain Tissue

1. Human fetal brain is obtained from Advanced Bioscience Resources (ABR), Inc., Alameda, CA 94501, USA in accordance with NIH guidelines.
2. One timed-pregnant, embryonic day 16 or 17 (ED16 or ED17) Sprague-Dawley rat (Charles River Laboratories) (*see* **Note 1**).

2.2 Reagents

1. Tryple™ Express (Gibco, cat# 12605) (*see* **Note 2**) plus DNase I (10 U/ml).
2. Immunopanning cocktail: 50 mM Tris–HCl, pH 9.5, 1 % BSA (w/v), 50 μg/ml anti-PSA-NCAM antibody (Developmental Studies Hybridoma Bank, University of Iowa, DSHB, 5A5).
3. Neuronal medium: Neurobasal medium, B27 supplemented with vitamin A, 0.25 mM GlutaMAX (Invitrogen), 5 ng/ml BDNF 100 U/ml penicillin and 100 U/ml streptomycin.
4. 70 % ethanol.
5. 4 % paraformaldehyde (PFA) in 1× PBS.

2.3 Dissection Instruments, Plasticware, and Glassware (All Sterile)

1. Dissecting forceps, straight.
2. Dissecting forceps, blunt.
3. Mayo dissecting scissors 5.5 in., straight.
4. Iris scissors, straight.
5. Iris scissors, curved.
6. Delicate dissecting scissors, straight.
7. Spatula, microdissecting.
8. Handi-Hold* Microspatula.
9. Cell strainers, mesh size 70 μm.
10. 60 mm tissue dishes.
11. 100 mm petri dishes (bacterial grade, non TC-treated).
12. Conical tubes 15 and 50 ml.
13. Glass slides.

14. Cover slips.
15. 37 °C, 5.0 % CO_2 incubator, 95 % humidity.
16. Biosafety laminar flow/tissue culture hood.

3 Methods

We adjusted protocols provided by the Millipore to our laboratory settings and developed additional steps to insure successful results while using the AXIS device.

3.1 Preparation of Slides

This is the one of the most important steps in this protocol. Perform all steps in a tissue culture hood in sterile conditions.

1. Soak the glass slides in 70 % ethanol for at least for 3 h in sterile petri dishes a day before preparation of cells. Soak forceps in 70 % ethanol at least for 3 h in sterile glass beaker a day before preparation of cells.
2. Wash slides with sterile water three times.
3. Aspirate excess of water and air-dry slides.
4. Transfer a single air-dried slide into new petri dish using sterile forceps.
5. Pipette 500 μl of 0.5 mg/ml poly-D-lysine solution on the slide and place a second sterile air-dried slide onto it. Incubate overnight at 4 °C. Both glass slides will be coated with poly-D-lysine.
6. Next day gently separate slides using two forceps, rinse with sterile water and air-dry at least for 2 h.
7. Place the poly-D-lysine coated slide in new sterile petri dish (coated surface is facing up). The slide is ready to use.

3.2 Assembling the AXIS™ Device on Slides and Preparation for Plating of Cells

AXIS™ Axon Isolation Devices (Millipore) enable easy isolation, observation and testing of developing neurites. It consists of four wells, two channels, and a set of microgrooves. Two of the wells are interconnected by a channel on each side of the device. The microgrooves are located in the area between the two channels (Fig. 1).

All steps should be performed in sterile conditions in the tissue culture hood.

1. Using sterile forceps carefully place AXIS device on the coated surface of the glass slide with the microgroove imprinted side facing down. The AXIS device will adhere easily to the glass. The space will be formed between imprinted surface on the AXIS and the glass slide surface. Device can be carefully repositioned if necessary. It is important to avoid touching

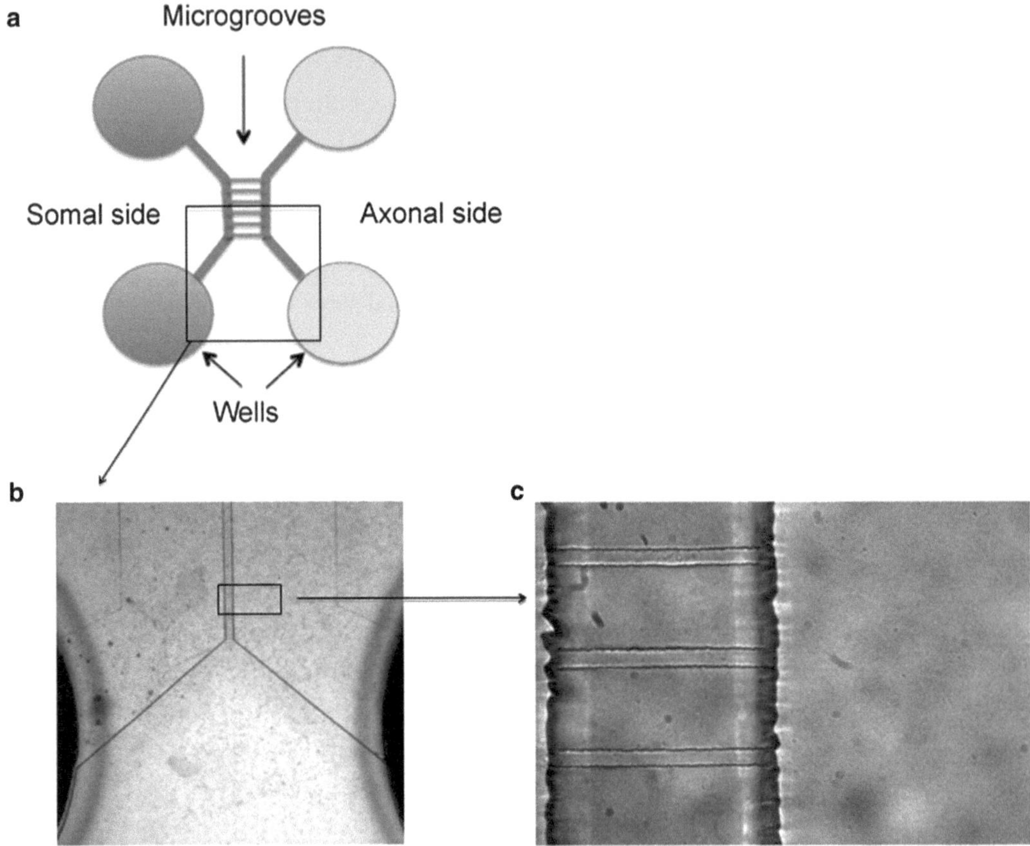

Fig. 1 (**a**) Schematic presentation of AXIS device. (**b**) Phase image of the portion of the assembled device (the *black box* on panel **a**), containing the microgrooves and adjacent chambers connected with wells. (**c**) Phase image of the microgrooves

imprinted surface to prevent damage to microgrooves and preserve the structural integrity of the device. Designate 2 wells (A and B) for applying cell suspension. Other two (C and D) will be cell free (*see* **Note 3**).

2. Apply gentle pressure to the top surface of the device at various points using blunt forceps. Avoid excessive force, which can compress and deform channels.
3. Add 200 μl of neuronal media to well A. Avoid bubbles, which can enter the channel and block it. Media should flow through the channel connecting A and B wells and will appear at the ostium of the channel in the well B quickly (during 1–2 min). If there is a delay, set the pipette at 100 μl and pipette up and down 3–4 times with the pipet tip at the ostium of the channel in well A. Avoid bubbles.
4. Add 200 μl of neuronal media to well B. Incubate 15 min at room temperature.

Fig. 2 (**a**) Neurons immunostained with anti-βIII tubulin antibody (*green* fluorescence). (**b**) DAPI-labeled nuclei (*blue*) visualized in the somal side chamber. (**c**) Merged image

5. Add 100 μl of neuronal media to well C. Media should flow through the channel connecting C and D wells and will appear at the ostium of the channel in the well D.
6. Add 100 μl of neuronal media to well D.

3.3 Plating Neurons

1. Count cells using hemacytometer. Prepare 500 μl cell suspension at a concentration 5–6 million cells per ml in neuronal medium (with 2 ng/ml BDNF).
2. Aspirate the media from all wells.
3. Plate 7 μl of cell suspension directly into the channel connecting wells A and B and incubate AXIS device with cells for 20 min in tissue culture incubator at 37 °C (*see* **Note 4**).
4. Add 100 μl of neuronal media to wells A and B. Add additional 100 μl of media to wells A and B (total 200 μl).
5. Add 100 μl of neuronal media (with 10 ng/ml BDNF) to wells C and D. The BDNF concentration gradient between axonal and somal side stimulate directed growth of neurites.
6. Incubate AXIS device containing neuronal cultures in tissue culture incubator at 37 °C.
7. Monitor cells closely, add fresh media every second day.

The process of extension into the axonal side chamber can be assessed by immunocytochemistry using anti-βIII tubulin antibody after 10–12 days in vitro (Fig. 2).

Recent studies have demonstrated that axons similar to dendrites, are capable of synthesizing new proteins via translation of localized mRNAs [2–6]. The importance of local protein synthesis has been established in regenerating axons [5, 7, 8]. Using the AXIS device it is possible to identify the potential roles of proteins and RNA transcripts present in neuronal processes exposed to various conditions [9]. The axonal mRNA or protein pool can be compared to total neuronal RNA or protein pool [1, 10, 11], which can help to obtain important information regarding molecular processes in neurites in response to signals.

4 Notes

1. Protocols using human tissue and animals must be approved by an Institutional Animal Care and Use Committee (IACUC).
2. Tryple™ Express is a special formulation of Trypsin/EDTA optimized for the culture of neurons.
3. When using several devices in one experiment it may save time and can be convenient to place all devices on the slides in the same orientation.
4. If necessary, increase the incubation time to avoid displacement of cells. Do not disturb cells and keep AXIS device horizontal to prevent passage of cells through microgrooves to the axonal side.

References

1. Park JW, Vahidi B, Taylor AM et al (2006) Microfluidic culture platform for neuroscience research. Nat Protoc 1:2128–2136
2. Taylor AM, Berchtold NC, Perreau VM et al (2009) Axonal mRNA in uninjured and regenerating cortical mammalian axons. J Neurosci 29:4697–4707
3. Sutton MA, Schuman EM (2006) Dendritic protein synthesis, synaptic plasticity, and memory. Cell 127:49–58
4. Campbell DS, Holt CE (2001) Chemotropic responses of retinal growth cones mediated by rapid local protein synthesis and degradation. Neuron 32:1013–1026
5. Wu KY, Hengst U, Cox LJ et al (2005) Local translation of RhoA regulates growth cone collapse. Nature 436:1020–1024
6. Lin CH, Forscher P (1993) Cytoskeletal remodeling during growth cone-target interactions. J Cell Biol 121:1369–1383
7. Willis DE, Twiss JL (2006) The evolving roles of axonally synthesized proteins in regeneration. Curr Opin Neurobiol 16:111–118
8. Yaron A, Zheng B (2007) Navigating their way to the clinic: emerging roles for axon guidance molecules in neurological disorders and injury. Dev Neurobiol 67:1216–1231
9. Quinn CC, Wadsworth WG (2008) Axon guidance: asymmetric signaling orients polarized outgrowth. Trends Cell Biol 18:597–603
10. Taylor AM, Blurton-Jones M, Rhee SW et al (2005) A microfluidic culture platform for CNS axonal injury, regeneration and transport. Nat Methods 2:599–605
11. Chilton JK (2006) Molecular mechanisms of axon guidance. Dev Biol 292:13–24

Chapter 14

Quantitative Assessment of Neurite Outgrowth Over Growth Promoting or Inhibitory Substrates

George M. Smith, Yingpeng Liu, and Jee W. Hong

Abstract

The use of sensory neurons and assessment of neurite outgrowth in vitro is an important part of understanding neuronal development and plasticity. Cultures of rat dorsal root ganglion (DRG) neurons provide quantitative results very quickly and, when grown on growth promoting or inhibitory substrates, can be utilized to study axonal growth, neurotrophic dependence, structure and function of growth cones. Since we are interested in axon regeneration and targeting, we have sought to promote neurite outgrowth by refining the techniques of growing DRG neurons in culture. This chapter describes detailed methods for the dissection and purification of DRG neurons and quantitative assessment of neurite on promoting or inhibitory substrates.

Key words Dorsal root ganglia (DRG) neurons, Axonal growth, Promoting or inhibitory substrates, Neurite quantitation, Image analysis

1 Introduction

The adult mammalian CNS is extremely limited in its ability to regenerate axons following injury. Failure of axons to regenerate after injury results in permanent functional deficits. Recent studies in different species and using different injury models have discovered important cellular and molecular mechanisms that govern axon regeneration. A number of molecules contribute to the failure of neural regeneration in CNS neurons and are associated with myelin, neuroinflammatory processes, glial scar, and CNS neurons themselves [1–3]. Such molecules act within signaling pathways which have inhibitory influences both extrinsically in the environment and intrinsically impacting neuronal growth [4–6]. Previous studies focused on the ability of neurons to regenerate, inhibitors of axon regeneration and strategies to overcome this inhibition utilize both in vitro and in vivo approaches [7–10]. Animal models of CNS injury are commonly used to examine the injury environment and axon regeneration, but animal studies are usually not suitable

Shohreh Amini and Martyn K. White (eds.), *Neuronal Cell Culture: Methods and Protocols*, Methods in Molecular Biology, vol. 1078, DOI 10.1007/978-1-62703-640-5_14, © Springer Science+Business Media New York 2013

for initial screening and characterization of therapeutic factors. Neuronal cell culture studies can provide a faster and easier way for examining the molecular components of neural regeneration. Though cultures from a variety of neuronal sources have been used for this purpose, dissociated dorsal root ganglion (DRG) neurons from postnatal rodents are among the most widely used of neuronal cell types. These DRG cultures can be maintained without trophic factors in Neurobasal A medium. Sensory neurons grow over growth promoting or inhibitory substrates have been a useful tool in studies for (1) assessment of neurite outgrowth on different substrates, (2) identification of molecules that can promote or inhibit neurite outgrowth, and (3) identification of compounds that can overcome the inhibitory effects of repulsive substrates.

Here, we demonstrate a method that employs a density gradient which can selectively purify and yield high quantities of DRG neurons over other cell types. The neurons can be plated onto growth promoting or inhibitory substrates. After the cultured cells are fixed and immunolabeled, microscope images are captured to ascertain neurite outgrowth. Quantitative measurements can be made to determine number of neurons, total neurite length, branch number and length. We have used the method described here to study the effects of growth factors and different signaling pathway modulators on neurons. In this chapter, we provide a detailed method for isolation of mammalian dorsal root ganglion neurons for use in culture and quantitative assessment of neurite.

2 Materials

2.1 Reagents

1. 70 % ethanol.
2. 15 mm round coverslips.
3. Poly-L-Lysine.
4. Laminin.
5. CSPG, a mixture including neurocan, phosphacan, versican, and aggrecan (Chemicon).
6. Hanks' Balanced Salt Solution (HBSS).
7. Collagenase type I in HBSS (store in 1.5 ml aliquots at –80 °C).
8. 0.25 % Trypsin/EDTA.
9. FBS.
10. DMEM.
11. Neurobasal A medium.
12. B-27 supplement.
13. GlutaMAX.

14. Penicillin/streptomycin.
15. Percoll, sterile (GE Healthcare).
16. Mouse βIII tubulin (Promega).
17. Goat anti-mouse Tex Red (Jackson Laboratories).
18. Fluoromount G.
19. Surgical instruments for dissection of DRGs include large, small scissors, # 5 dissecting forceps.
20. Horizontal laminar-flow tissue culture hood.

2.2 Neurobasal A Based Culture Medium

24 ml Neurobasal A medium.

500 μl 50× B-27 supplement.

250 μl 100× Glutmax.

250 μl 100× penicillin/streptomycin (10,000 U/ml each).

Prepare fresh medium each day before use (*see* **Note 12**).

2.3 Animals

The procedures described below are optimized for culture of DRG neurons from a 3 to 6 week-old Sprague-Dawley rat (Harlan Bioproducts for Science) (*see* **Note 1**). All animal experiments must be approved by an Institutional Animal Care and Use Committee (IACUC) and must follow guidelines and regulations for the care and use of laboratory animals.

2.4 Preparation of Coverslips

1. Place glass coverslips into a rack and submerge in 65 % nitric acid overnight.
2. Wash coverslips in sterile water three times for 30 min.
3. Put racks with coverslips in 70 % ethanol overnight.
4. Place coverslips into an appropriate culture dish (one slide per well of a 12-well plate).

2.5 Preparation of Permissive Substrate

1. Make a stock solution of 20 mg/ml Poly-D-Lysine (PDL) diluting with tissue culture-grade water and store in aliquots at −80 °C.
2. The PDL stock solution is thawed and diluted to a final concentration of 0.1 mg/ml.
3. Coat plates with the following volumes: 300 μl/well for 4 well chamber slides; 500 μl/well for 24 well plates and 1 ml/well for 12 well plates.
4. Add PDL solution to cover the well and incubate coverslips in a humidified 37 °C/5 % CO_2 incubator for at least 2 h.
5. Remove the PDL solution by vacuum and wash four times with sterile ddH_2O (*see* **Note 5**).
6. Allow coverslips to dry completely in a laminar flow hood overnight.

7. To study axon growth on permissive substrates, glass coverslips are further coated with laminin (2–10 μg/ml) in a humidified 37 °C/5 % CO_2 incubator for at least 1 h (*see* **Note 4**).
8. Before plating cells, aspirate the laminin solution from the coated wells of a PDL and laminin-coated plate (*see* **Note** 7).

2.6 Preparation of Inhibitory Substrate

1. For preparing inhibitory substrates, glass coverslips are coated with PDL (100 μg/ml) in a humidified 37 °C/5 % CO2 incubator for at least 2 h.
2. Wash four times with sterile ddH_2O and dry overnight.
3. Add a mixture of 5 μg/ml laminin and 10 μg/ml CSPG to a PDL-coated plate and incubate overnight at 4 °C.
4. Wells are washed two times with HBSS before plating cells.

2.7 Percoll Gradient Preparation

After trituration of cells, prepare Percoll gradient to separate myelin, glial cells and nerve debris from DRG neurons. Add 1.1 ml Percoll to 2.9 ml HBSS in one 15 ml tube (28 % Percoll) and add 0.5 ml Percoll to 3.5 ml PBS in another tube (12 % Percoll). Carefully layer the 12.5 % Percoll onto the top of the 28 % Percoll solution, with minimal disturbance to the lower Percoll layer. *A clear colorless solution should be seen on the top of a layer of red colored solution* (Fig. 1c). A boundary should be visible between the two layers.

3 Methods

3.1 Dissection and Dissociation

1. Sterilize all dissection instruments by autoclave for at least 20 min before using.
2. Euthanize the rat by CO_2.
3. Spray the rat whole body, especial the dorsal area, with 70 % ethanol.
4. Move the rat to the dissection area in the laminar flow hood.
5. Decapitate the rat by cervical transection.
6. Using large scissors, cut the skin along middle of back from neck to tail and pull the skin to expose cervical, thoracic and lumbar spinal regions (*see* **Note 2**).
7. Open the dorsal side of the spinal column to exposing the spinal cord using a small scissors.
8. Under the dissecting microscope, cut through both sides of dorsal spinal column and remove long strip of spinal column.
9. Remove the spinal cord by cutting central sensory roots from the animal to expose DRGs between the vertebrate.
10. Dissect out the DRGs and cut off peripheral sensory roots.

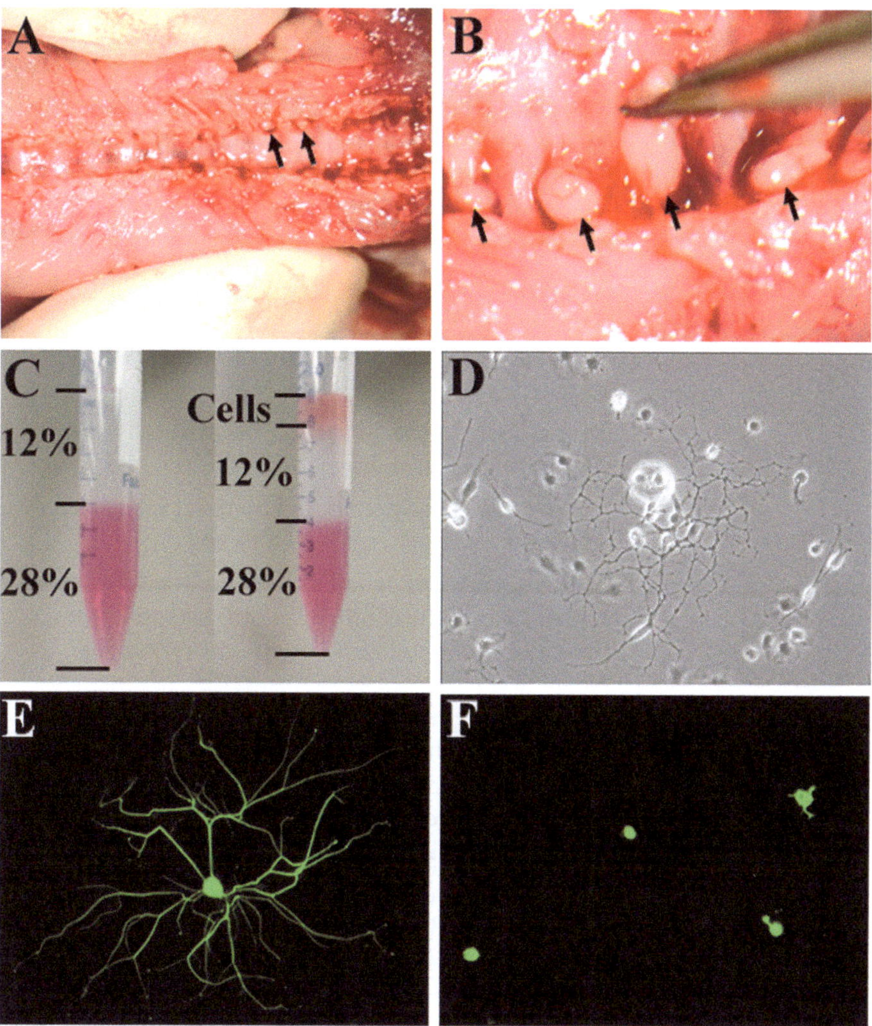

Fig. 1 (**a**) After removal of the dorsal part of the spinal column and spinal cord, the ventral spinal column can be identified and DRGs attached on both sides. *Arrows* indicate the location of DRGs within the lateral wall of the vertebral column. (**b**) DRGs as seen under the microscope at higher magnification are located within the intervertebral foramen (*arrows* point to several DRGs, one of them pinched by forceps). DRGs are identified by their grape-like shape. (**c**) To eliminate the myelin debris and decrease the non-neuronal cells, the dissociated cell suspension was layered onto 8 ml of freshly made 12 %/28 % Percoll solution. *Left panel* shows the top layer of 12 % Percoll (*colorless*) and the bottom layer of 28 % Percoll (*red*). *Right panel* shows the cell suspension composed of neurons, non-neuronal cells, and myelin debris at the top of the gradient solution before Percoll centrifugation. (**d**) Phase-contrast micrographs of primary DRG neurons in culture after isolation from a 4-week-old rat were plated onto a Laminin coated substrate and grown for 18 h. After Percoll centrifugation, most of the myelin debris has been eliminated and some non-neuronal cells still existed. (**e** and **f**) DRG neurons cultured for 18 h on a substrate with permissive substrate (Poly-l-Lysine/Laminin) or inhibitory substrate (Poly-l-Lysine/Laminin/CSPG). DRG neurons were fixed and visualized with an antibody to the high-molecular weight neurofilament. Neurons develop longer axonal arborizations on the laminin substrate (**e**) or short axon stumps on CSPG substrate (**f**)

11. The ganglia are carefully freed of connective tissue, and the roots are cut close to the ganglion (Fig. 1a and b).
12. Place the DRGs in HBSS media until dissection is complete.
13. Carefully aspirate the HBSS medium using a hand operated pipette to avoid vacuum aspiration of the DRGs, add 1 ml of 0.25 % collagenase solution and incubate 45–60 min at 37 °C (*see* **Note 8**).
14. Collagenase is thereafter replaced by 1.5 ml 0.25 % trypsin/EDTA and incubated at 37 °C for 15 min.
15. Ganglia are washed with 1.5 ml of 10 %FBS-DMEM to terminate digestion.
16. Use a polished glass pippet to mechanically dissociate DRGs by aspirating up and down (*see* **Note 6**).
17. After five cycles, undissociated material is allowed to sediment and the supernatant removed and saved.
18. The cell suspension is filtered through a 100 μm nylon mesh. This step allows the elimination of substantial amount of myelin debris and non-dissociated fragments of ganglia.

3.2 Separation of Cells by Density Gradient Centrifugation

1. To discard the myelin debris and purify the dorsal root ganglion neurons, the dissociated cells are layered onto the top of a 12.5/28 % Percoll solution and centrifuged at 1,300×*g* for 10 min (*see* **Note 10**).
2. Discard the top 5 ml of supernatant including interface (most of the neurons are in the 28 % Percoll layer) and dilute to a final volume of 8 ml with HBSS (Fig. 1c).
3. Centrifuge at 1,000×*g* for 6 min and discard the supernatant.
4. Resuspend the pellet in Neurobasal A based culture medium and count cells with a hemacytometer (*see* **Note 9**).
5. Dilute the cell suspension in Neurobasal medium at an appropriate density depending on the application (*see* **Note 11**).
6. Add cell suspension at desired cell density to each well (*see* **Notes 3** and **13**).
7. Place the labeled multi-well plate into a humidified 37 °C, 5 % CO_2 incubator and let the cells sit undisturbed for several hours to allow attachment (Fig. 1d).
8. Exchange medium every 2 days by changing 50 % medium (*see* **Note 14**).

3.3 Immunocytochemical Staining

To visualize the DRG neurons with axons for neurite outgrowth assay, the cultured cells are fixed in 4 % paraformaldehyde in PBS for 25 min and washed four times with PBS. The cells are exposed to the primary antibodies: mouse anti-βIII tubulin(1:1,000) for

2 h at room temperature or 4 °C overnight. The cultures are washed with PBS three times for 5 min each. Secondary antibodies: goat anti-mouse Tex Red (1:200) are applied for 1 h at room temperature and then washed.

3.4 Image Capture

In order to photograph the DRG neurons, each cover glass is placed on the microscope slide using Fluoromount G and allowed to settle for 5 min. Randomly identified single neurons are photographed using Nikon Eclipse 80i microscope equipped with Nikon Plan Fluor 10×/0.30 objective. Neurons stained with goat anti-mouse Tex Red are detected using Nikon Intensilight C-HGFI filter. Images are captured with Nikon DS-Ri1 camera and processed through Nikon Digital Sight. The ideal image captured will show just one individual neuron.

3.5 Measurements of Neurite Length

NIS-Element BR 3.0 software is used to measure neurite length of each image. Exposure time for images is appropriately adjusted one at a time. In the software, Polyline under annotations and measurements is used to trace neurite length and measurements are automatically recorded. The recorded neurite length of each image is then transferred to Microsoft excel program for statistical analysis. Each image is referred back as the neurite lengths are added for each cell in the excel sheet (*see* **Note 15**).

4 Notes

1. Neurons cultured from early postnatal rats (<7 days old) grow axons faster and longer than neurons from older adult rats. Neurons from adult rats may have shorter axons and be more difficult to maintain in culture. Adjust the substrate to keep appropriate axon lengths in accordance with the age of the animal and the type of experiment.
2. Before opening up the spinal column, make two cuts on either side to remove a thin tendon strip. Make sure not to cut too laterally along the sides, because DRGs are located within the lateral vertebral wall. However in order to see DRGs it is necessary to remove the dorsal bones from the spinal column.
3. For the best growth of DRG neurons and neurite assay, make sure all dishes or coverslips are well coated with promoting or inhibitory substrates.
4. It is important to keep laminin on ice thawing it slowly before coating.
5. Unbound PDL is detrimental to cells; therefore it is important to rinse coverslips or dishes thoroughly with sterile ddH_2O and dry them overnight in the hood.

6. Over-trituration, over-digestion or wrong pH of medium can prevent cells from adhering to the dish and cause them not to extend processes. Under-trituration or shorter digestion time can lead to cell clumping. It is important to optimize and find appropriate times for trituration and digestion.
7. PDL coated plates are stable for up to 2 weeks. Laminin and/or CSPG coated plates can be stored at 4 °C for 1 week. For good cell growth, however, we use freshly coated plates every time.
8. DRGs obtained from young animals require less enzymatic digestion and milder trituration. Cultures of DRG neurons from older rats (older than 6 months) require longer digestion and more aggressive dissociation. Therefore, it is important to adjust the procedures for optimized culture conditions according to different ages of animals.
9. For culture medium, we find it is best to prepare fresh each time.
10. Some studies target only the intrinsic growth capacity of neurons. In order to avoid interference of trophic activities secreted by non-neuronal cells, it is essential to separate neuronal populations as much as possible from non-neuronal cells.
11. Neurite outgrowth from dissociated DRG neurons is affected by low cell density and low laminin or high CSPG concentrations. It is better to verify the cell density and substrate concentration are appropriate. Also low-density plating is important for certain experiments since it minimizes interactions between neurons and non-neuronal cells and allows for more accurate neuronal measurements.
12. Purified adult DRG neurons without non-neuronal cells require trophic support for their short-term survival and attachment. Addition of conditioned medium from Schwann cells or astrocytes at beginning will help neurons to attach to the plate.
13. Maintain consistent cell density in every well by mixing the cell suspension while plating and agitate the plate to distribute the cells evenly.
14. Replace half of the medium every other day and use gentle technique when changing medium so as not to disturb attached neurons.
15. Inhibitory substrates can be generated to form alternating stripes to examine axon behavior upon contact with an inhibitory substrate [11], instead of uniform distribution as described in this method.

Acknowledgment

This work was supported by NIH grant RO1 NS060784, Shriners Hospitals for Children (GMS) and Shriners Postdoctoral Fellowship (G.M.S. and Y.L.).

References

1. Fitch MT, Silver J (2008) CNS injury, glial scars, and inflammation: inhibitory extracellular matrices and regeneration failure. Exp Neurol 209:294–301
2. Yiu G, He Z (2006) Glial inhibition of CNS axon regeneration. Nat Rev Neurosci 7:617–627
3. Sun F, He Z (2010) Neuronal intrinsic barriers for axon regeneration in the adult CNS. Curr Opin Neurobiol 20:510–518
4. Filbin MT (2003) Myelin-associated inhibitors of axonal regeneration in the adult mammalian CNS. Nat Rev Neurosci 4:703–713
5. Liu K, Lu Y, Lee JK, Samara R, Willenberg R, Sears-Kraxberger I et al (2010) PTEN deletion enhances the regenerative ability of adult corticospinal neurons. Nat Neurosci 13:1075–1081
6. Qiu J, Cai D, Dai H, McAtee M, Hoffman PN, Bregman BS et al (2002) Spinal axon regeneration induced by elevation of cyclic AMP. Neuron 34:895–903
7. Neumann S, Bradke F, Tessier-Lavigne M, Basbaum AI (2002) Regeneration of sensory axons within the injured spinal cord induced by intraganglionic cAMP elevation. Neuron 34:885–893
8. Park KK, Liu K, Hu Y, Smith PD, Wang C, Cai B et al (2008) Promoting axon regeneration in the adult CNS by modulation of the PTEN/mTOR pathway. Science 322: 963–966
9. Moore DL, Blackmore MG, Hu Y, Kaestner KH, Bixby JL, Lemmon VP et al (2009) KLF family members regulate intrinsic axon regeneration ability. Science 326:298–301
10. Qiu J, Cafferty WB, McMahon SB, Thompson SW (2005) Conditioning injury-induced spinal axon regeneration requires signal transducer and activator of transcription 3 activation. J Neurosci 25:1645–1653
11. Snow DM, Smith JD, Cunningham AT, McFarlin J, Goshorn EC (2003) Neurite elongation on chondroitin sulfate proteoglycans is characterized by axonal fasciculation. Exp Neurol 182:310–321

Index

A

B

C

D

E

F

G

H

Shohreh Amini and Martyn K. White (eds.), *Neuronal Cell Culture: Methods and Protocols*, Methods in Molecular Biology, vol. 1078, DOI 10.1007/978-1-62703-640-5, © Springer Science+Business Media New York 2013

MIX
Papier aus verantwortungsvollen Quellen
Paper from responsible sources
FSC® C105338

If you have any concerns about our products,
you can contact us on
ProductSafety@springernature.com

In case Publisher is established outside the EU,
the EU authorized representative is:
Springer Nature Customer Service Center GmbH
Europaplatz 3, 69115 Heidelberg, Germany

Printed by Libri Plureos GmbH
in Hamburg, Germany